9/28/94

D1480570

Ecological Communication

Ecological
Communication

NIKLAS LUHMANN
Translated by John Bednarz, Jr.

The University of Chicago Press

Originally published as *Ökologische Kommunikation: Kann die moderne
Gesellschaft sich auf ökologische Gefährdungen einstellen?*
© 1986 Westdeutsches Verlag GmbH, Opladen.

The University of Chicago Press, Chicago 60637
Polity Press, Cambridge

© 1989 by Polity Press
All rights reserved. Published 1989
Printed in Great Britain

98 97 96 95 94 93 92 91 90 89   54321

Library of Congress Cataloging-in-Publication Data

Luhmann, Niklas.
  Ecological communication.
  Translation of: Ökologische Kommunikation.
  Includes index.
  1. Ecology—Philosophy. 2. Man—Influence on nature.
3. Environmental protection—Philosophy. 4. Environ-
mental policy. I. Title.
QH540.5.L8313 1989   574.5′01   89–4843
ISBN 0–226–49651–1 (alk. paper)

This book is printed on acid-free paper.

Contents

Translator's Introduction

Ecological Communication enjoys a unique position in the extensive writings of Niklas Luhmann. As a late work it embodies the most recent developments in a thought that has undergone important changes. It is relatively short but also includes every aspect of Luhmann's complex theoretical position, and it provides the theoretical framework within which concrete social themes and problems are addressed and handled. For these reasons *Ecological Communication* is an excellent introduction to the current position of Luhmann's thought and an indication of its direction. But at the same time the aspects that make it such an attractive work also make it difficult to master.

Ecological Communication is very compact and demands an extensive background knowledge from its readers. The English-speaking audience is at an immediate disadvantage here because only a small fraction of Luhmann's writings are available in translation. (One German reviewer, for instance, estimates that there are at least 6,000 pages of Luhmann's work in print in German today.) But even for the German audience what was required for a complete presentation of his position was not available until the publication of his most comprehensive work, *Soziale Systeme* (1984).[1]

In an effort to make Luhmann's position more accessible I would like to make some general remarks to situate *Ecological Communication* within it and, I hope, provide a more adequate basis for its comprehension.

Luhmann's general position is a unique and extraordinarily

powerful synthesis of several quite diverse intellectual traditions. At least four of these can be distinguished: (1) the systems-theoretical approach to social action found in the writings of Talcott Parsons; (2) the cybernetic interpretation of the relationship between system and environment; (3) a phenomenological disclosure of meaning and its importance for the relationship of the components of social systems; and (4) an autopoietic understanding of system-organization. *Ecological Communication* – and indeed all Luhmann's recent works – can be understood only through a clarification of these four separate intellectual traditions and the way in which he combines them. I would like to consider these separately to examine what advantage Luhmann draws from each and how this is exploited in *Ecological Communication*.

It is well known, certainly at least since *Theorie der Gesellschaft oder Sozialtechnologie: Was leistet die Systemforschung?* (1971),[2] that Luhmann's plan for a unified social theory is to be understood as a theory of social systems. Furthermore, this theory makes a claim to universality – and hence the reproach of ideology that is so often leveled against it. But in this case Luhmann makes clear that universality does not represent a claim to the exclusive possession of truth *vis-à-vis* other, competing theories. Instead, he means that this theory includes every aspect of the social domain within itself and not just certain ones like stratification and mobility, conflict and conformity, models of interaction, etc.

The influence of Talcott Parsons has been well documented in this respect and has already been submitted to detailed criticism. Its most lasting effect has been felt through the interpretation of social action as interaction. According to Parsons, interaction requires a plurality of actors whose actions are distinguished as a unity (system) *vis-à-vis* an environment through symbolically mediated structures of expectation.

This cursory presentation of a central aspect of Parsons's social theory reveals how important it is for Luhmann's general position because it focuses on the important concepts of complexity and contingency that are involved in all interaction. Complexity signifies the 'potential for high degrees of differentiation among the components which . . . constitute . . . systems',[3] while contingency, or rather double contingency, denotes the 'complementarity of expectations in the process of human interaction'.[4] Both these concepts assume a central place in Luhmann's theory of social systems too. But Luhmann emphasizes the integrative role of meaning.

The significance of Parsons's systems-theoretical approach for *Ecological Communication* is its interpretation of ecological problems from a general-systems perspective, i.e., in terms of the relation of a system – in this case society – and its environment. When ecological problems are approached in this way the concepts and methods of systems theory – especially its most recent developments – can be exploited not only to reveal the conditions for these problems but also ways in which their resolution becomes a possibility.

The concept of complexity employed by Parsons as an integral part of his systems theory of social action is also a corner-stone of another systems-theoretical discipline. Cybernetics is the latest development in the history of systems-thinking starting with Aristotle. It interprets the relation between system and environment as a difference in degree of complexity. Because a system's environment includes everything other than the system itself, its complexity is always much greater than that of the latter. This means that systems are constantly confronted with new and different environmental states. To deal with these they have to bring their own complexity into a relation of correspondence with that of their environments. Systems do this through establishing system structures that reduce the complexity of their environments and thereby obviate point-for-point correlations between their own changes and changes in their environments. The ecological significance of complexity appears as the problem of a societal environment that can always change in more ways than society itself. The latter, however, still has to react to these changes. In *Ecological Communication* Luhmann uses the concept of 'resonance' to designate this system/environment interplay. But while complexity – environmental complexity – is always a problem for a system, at the same time it is the key to its solution through the increase of the system's own complexity.

Complexity, which is both the central problem in a cybernetic systems-theory as well as the key to its solution, means that system components are related, connected together. Social and psychical systems are unique because this connection is accomplished through meaning. Now the concept of meaning is nothing new to the investigation of the social domain. Parsons for instance, following Weber, interpreted meaning as a subjective *property* of actions in order to overcome a purely behavioristic standpoint. Luhmann, however, understands it as a determinate strategy amongst alternative possibilities. In other words, he no longer understands meaning substantively as a *property* of system components but functionally as the constituting and integrative *relations* among them. This is the point at which Luhmann resorts to phenomenological descriptions because, according to him, only the latter can reveal meaning.[5] Only a phenomenological description of experience can reveal how the elements of social systems refer to – are connected with – others, especially in the past and future. It also reveals how the others are inherently contingent, i.e. not necessary. As with Parsons, social action is possible only when the meaningful perspectives of different actors can be brought into agreement. System structures accomplish this by reducing the complexity inherent in all social action.

The significance of meaning for *Ecological Communication* is the essential recursiveness (self-referentiality) of system components – in this case communications. Phenomenological investigations· reveal that meanings constitute themselves self-referentially, i.e., they refer exclusively to other meanings. They thereby organize what Luhmann – following Husserl – calls 'horizons' of further communicative alternatives. Communication can communicate only what is meaningful because for Luhmann it is not a 'transfer' of information but instead the common actualization of meaning.

The concept of meaning leads directly into the fourth and final aspect of Luhmann's complicated position: the concept of autopoiesis. This concept was first introduced by the theoretical biologists Humberto Maturana and Francisco Varela in *De Maquinas y Seres Vivos* in 1972 (*Autopoiesis and Cognition*, 1980). They used it to capture the unique capacity of living systems to maintain their autonomy and unity through their very own operations. Maturana and Varela discovered that this is possible only when the operations proper to a system (its complex of component-producing processes) and its components (elements) themselves are so related that they constantly (re)produce each other. They expressed this relation as the organizational closure of the system. By distinguishing systematically between system organization and system structure, a model was established for a dynamic system that was both closed (organizationally) and (structurally) open at the same time.

Despite the initial difficulties encountered by Maturana and Varela in extending this model beyond the biological domain, Luhmann exploits both its analytic and explanatory power by applying it to social systems where it also appears under the name of self-reference. At the basic level this means that social systems constitute themselves self-referentially. Everything that functions as an element in the system is itself a product of the system. Luhmann reveals how this occurs by carefully distinguishing between (system) element and (system) relation, i.e., structures and processes. He shows that the concept of autopoiesis can be extended to the social domain only when the elements of social systems are conceived as communicative acts (events) and not as persons, roles, subjects, individuals, etc.

The importance and uniqueness of the concept of autopoiesis for Luhmann resides in its ability to provide the theoretical framework within which social systems can be differentiated as constituting themselves self-referentially through the development of their own separate symbolically generalized media of communication, for example, money/economy, power/politics, love/family and truth/science. Yet, unlike the case of earlier systems-theories where the system was either open or closed but not both, autopoiesis accounts at the same time for the non-reductive relation between the system and its necessary infrastructure (environment). In this sense, whatever is not communication is environment for society; even when this includes the consciousnesses – psychical systems – and lived bodies – biological systems – of human beings. After all, communication needs both psychical and biological systems in order to occur.

Not until the concepts of autonomy (closure) and interpenetration (openness) are unified in one theory can any systems theory as such, and any theory of social systems in particular, be complete. By doing this the concept of autopoiesis provides the synthetic unity necessary for the production of a systems theory of the social domain.

The importance of autopoiesis for *Ecological Communication* is the recursive organization of all social systems. When society is defined as the social system that includes the possibility of communication as such, this means that it cannot communicate with its environment and ecological problems can only be handled from within society itself, i.e., through its subsystems.

The synthesis of these four diverse intellectual traditions is certainly the source of the great strength and innovation of Luhmann's position in general and of *Ecological Communication* in particular. But at the same time it proves to be one of the greatest obstacles to their comprehension because all of these traditions developed separately and without any direct contact, for example, between phenomenology and cybernetics on one hand, or between the systems approach of Talcott Parsons and autopoiesis on the other. But even between those that enjoy a kind of conceptual relatedness – like cybernetics and autopoiesis – there is, to a significant degree, perhaps only a one-sided awareness. Therefore, one cannot assume that a reader of *Ecological Communication* – or any of Luhmann's recent works for that matter – who is familiar with one or even two of these traditions will also be familiar with the others and therefore able to appreciate the full import of what Luhmann presents. This becomes particularly evident in *Ecological Communication* where Luhmann applies his theoretical principles to an extraordinarily wide variety of social topics, ranging from law, politics, religion and education to science and the economy.

As the most recent addition to Luhmann's general theoretical position we have to start from the concept of autopoiesis in order to analyze and explain the central argument and themes of *Ecological Communication* because this is the concept that integrates, synthesizes the theoretical content of the other traditions. Autopoiesis signifies the closure of a system's organization, i.e., the self-reference of the complex of components and component-producing processes that mutually reproduce themselves and thereby integrate and unify the system. Organizational closure does not mean that a system cannot be affected at all by its environment. But it does mean that, as an autonomous unity, i.e., organizationally, it can react to its environment only in accordance with its own mode of operation, the mode of operation peculiar to it. Because society's specific mode of operation is communication it cannot react to its environment in any other way – and since its environment includes everything that does not operate in a communicative way, society is *eo ipso* prevented from communicating *with* its environment. Instead, it can only communicate *about* its environment *within* itself. In *Ecological Communication* the communicative reactions or disturbances that society's environment produces within society itself is called 'resonance'.

Luhmann believes that societal communication affects the very way in which the possibility of environmental dangers arise. So the question around which *Ecological Communication* turns is not one of how society can manage existing environmental problems. (Luhmann recognizes that this can be accomplished through a sufficiently powerful law for policing the environment.) Instead, he is concerned with how society comes to the very awareness of environmental dangers as such. How does it put itself in the position to recognize environmental dangers at all? After all, 'Ecological dangers may exist or not and no one may know about them. But the exposure to ecological dangers exists only when people communicate about the pollution of rivers and the air and the deforestation of the land.'[6] When this communication occurs, the dangers in the environment can be addressed only in the ways that society itself has established for communication. *Ecological Communication* is concerned with how these dangers shape up – it talks about the 'contours' of the problem of society's ecological adaption – within society when communication about them takes place. But this means that it is concerned only with 'how society, in fact, reacts to environmental problems and not how it ought to or would have to in order to improve its relation to the environment'.

Modern society, as opposed to those before it, is differentiated according to subsystems that concentrate on one specific and primary function. For this reason they are called 'function systems' in *Ecological Communication*. The economy, law, science, politics, religion and education are examples of these. Even if all the communication that occurs within society does not have to be ascribed to one of them, Luhmann argues that the socially most consequential communication does.

A particularly thorny problem, however, infects autopoietic systems. Self-reference means that their operations are applied to themselves. Ever since classical antiquity it has been known that this leads to paradox. (The classical example is the famous 'liar paradox' of Eubulides.) In recent times attempts to solve paradoxes of self-reference have produced the theory of types (Whitehead and Russell) and meta-languages (Tarski). These solutions involved the introduction of some form of *difference* – for example, the difference of types or linguistic levels – which, in Luhmann's terminology, 'interrupts' or 'unfolds' the (unity of the) self-reference. Function systems employ binary coding and programming to introduce this difference. This means that

> the *unity* (of self-reference, J.B.) that would be unacceptable in the form of a tautology (e.g., legal is legal) or a paradox (one does not have the legal right to maintain their legal right) is replaced by a *difference* (e.g., the difference of legal and illegal). Then the system can proceed according to this difference, oscillate within it and develop programs to regulate the ascription of the operations of the code's positions and counter-positions *without raising the question of the code's unity.*

Function systems structure their communication through binary codes that divide the world into two values (for example, true/false, legal/illegal, power/lack of power, immanence/transcendence, possessing/not possessing). The resonance created within society by its environment is channelled into one of these function systems and treated there effectively in accordance with one code or another. This complicates the ecological problem because from now on two system-references have to be kept separate: one between society as a whole and its environment (society's external boundary) and the other that exists within society itself for its particular function systems (society's internal boundaries). The problem is that environmental changes produce too little and too much resonance within society at the same time.

Coding and the programs that accompany it (theories in science, laws in the legal system, investments in the economy, party-political alignments in politics, etc.) produce a sharp reduction in what is information, i.e., differences that make a difference, for society. They screen it off from its environment organization-ally and obviate point-for-point correlations between system and environment. Resonance within society, then, is always improbable, i.e., occurs only in exceptional cases. According to Luhmann, this implies that society produces *too little resonance vis-à-vis* environmental dangers. On the other hand, the situation at the internal boundaries of society, i.e., where communication actually takes place, is quite different. Necessary communicative interdependencies among function systems can literally produce too much resonance within society *even if environmental disturb-ances produce too little resonance.*

What Luhmann has in mind here can be illustrated, for instance, by the relation between the economy and science or between politics and law. Furthermore, he also wants to indicate that insignificant changes in one of these function systems can trigger an 'effect-explosion' in another or others. For example, 'Theoretically insignificant scientific discoveries can have agoniz-ing medical results', and 'Payments of money to a politician that play no role in the economic process – measured by the billions of dollars that are transacted daily – can become a political scandal'. Again, the law can place 'the pharmaceutical industry and physicians under the threat of liability to supply information and to establish precautionary standards (that) can have medical as well as economic consequences that are entirely unrelated to what is legally important and might not even compromise part of the legal decision itself'.

The specific codes and programs of the particular function systems guarantee the disproportion in reactions to environmental disturbances among them. It also means that no one of the function systems can be substituted for another. Thus in the functionally differentiated modern society there is no subsystem (function system) that represents the whole of society from within as was the case in earlier societies – differentiated through stratification – when one level of society was viewed as the top or center of society.

The functional differentiation that produces a centerless society provides the basis for objections raised by *Ecological Communi-cation* against what it calls the 'new movements'. These use strategies that deal with environmental dangers by appealing to a new environmental morality. Accordingly, right is on the side of those who are against the self-destruction of society. A new environmental ethics for society is all that is necessary, and moral zeal – demonstrable through anxiety – can compensate for any theoretical deficits in this ethics.

But without a consideration of the constraints placed on ecological communication (including any environmental ethics) by the functional differentiation of society, this discussion can fall very easily into a rhetoric of anxiety.

Anxiety is a particularly attractive theme for ecological communication because it can always be used as the basis for moral justification when all else fails. But not much is gained through the communication (rhetoric) of anxiety except perhaps more communication (rhetoric) of anxiety, and although it can easily be used to block society's incursions upon its environment, it has to pay for this with unforeseen reactions on society that only produce more anxiety. If, then, a specific function is to be assigned to environmental ethics within the context of ecological communication, Luhmann says this should be to proceed cautiously in dealing with the morality of ecological problems.

John Bednarz Jr.

Preface

On 15 May 1985, at the invitation of the Rhenish-Westfallian Academy of Science, I addressed their yearly assembly on the theme, 'Can Modern Society Adjust Itself to the Exposure to Ecological Dangers?'[1] The restricted amount of time afforded by the address, however, did not allow for the presentation of what could be said on the matter in its entirety. Above all, it was impossible to deal with the very similar modes of reaction of the various particular function systems despite all their differences. The main argument of the address, namely that modern society creates too little as well as too much resonance because of its structural differentiation into different function systems, was presented only in its main outline. Only from this insight does it follow that the solution to this problem can be found in new ideas about values, a new morality or an academic elaboration of an environmental ethics.

The work presented here supplements the argument of my address and completes it for the most important function systems of modern society, even if only in very broad outlines. What emerges is a very similar picture of basic structure for the different functions, binary codes and programs of 'correct' experience and action. This justifies ascribing ecological problems to society and not only to the failures of politics and the economy or to an insufficient feeling of responsibility.

Perhaps in this way a theoretically guided comparison can clarify how social theory is challenged by ecological discussion that has recently become topical – and how little it had to offer formerly.

In many respects the analyses presented here diverge from the premises that have been introduced naively in the ecological literature and used without any further justification. This is true for basic systems-theoretical questions as well as for numerous details. Indeed, the ecological literature itself is a product of social communication, i.e., a part of the object that we are investigating here, including our own investigations. Indifference in the choice of words and absence of interest in significant theoretical decisions are two of the most noticeable characteristics of this literature – as if caring about the environment could justify carelessness in the talk about it. Moreover, it turns out that the literature that emerges with the claim to scientific precision is produced mainly by those disciplines that, at the same time, fulfill the reflective functions in those function systems from which they originate. This means that jurists must concern themselves with expanding the categories that deal with legal cases and economists with expanding those models with which economic data are observed and positive or negative economic growth is revealed. Of course, I recognize that all this has its significance. But the problem for me lies much deeper in the differentiation of the function systems themselves.

Investigations that are inspired theoretically can always be accused of a lack of 'practical reference'. They do not provide prescriptions for others to use. They observe practice and occasionally ask what is to be gained by making such a hasty use of incomplete ideas. This does not exclude the possibility that serviceable results can be attained in this way. But then the significance of the theory will always remain that a more controlled method of creating ideas can increase the probability of more serviceable results – above all, that it can reduce the probability of creating useless excitement.

Niklas Luhmann

1

Sociological Abstinence

Compared to the history of reflection on humanity and society this theme – ecology – is not very old. Only in the last twenty years has one seen a rapidly increasing discussion of the ecological conditions of social life and the connection between the social system and its environment. Contemporary society feels itself affected in many different ways by the changes that it has produced in its own environment. This is clearly shown by a number of these: the increasingly rapid consumption of non-replaceable resources and (even if this would prove beneficial) the increasing dependence on self-produced substitutes, a reduction in the variety of species forming the basis of further biological evolution, the ever-possible development of uncontrollable viruses resistant to medicine, the familiar problem of environmental pollution and not least of all over-population. Today these are all themes for social communication. Society has thus become alarmed as never before[1] without possessing, however, the cognitive means for predicting and directing action because it not only changes its environment but also undermines the conditions for its own continued existence. This is by no means a new problem. It appeared in earlier stages of social development too.[2] But only today has it reached an intensity that obtrudes as a 'noise' distorting human communication that can no longer be ignored.

As far as sociology is concerned, this discussion began – like so many others, unexpectedly – and caught it, as it were, unprepared theoretically. Originally, sociology had been con-

cerned with the internal aspects of society. It entangled itself in
ideologies of the correct social order and then tried to extricate
itself from them. All this was done under the assumption that
its theme was society or its parts. The history of the foundations
of this discipline had already predisposed it in this direction.
Nature, on the other hand, could and indeed had to be left to
the natural sciences. What the new discipline called sociology
could discover and claim as its own field of study was either
society or, if this concept was unsatisfactory, social facts, for
example, *faits sociaux* in Durkheim's sense, or social forms and
relations in the sense of Simmel and von Wiese, or social action
in Weber's sense. Thus the delimitation of the discipline had to
be interpreted as a demarcation of a section of reality.

But in addition to 'grand theory', research in the domain of
the most diverse *social problems* is also attuned to the social
origins of these problems. This is precisely what forms the basis
of the researcher's hopes of being able to contribute something
to a better solution of the problem. The problematic is reduced
to structures of the social system or its subsystems, and if these
cannot be changed then at least, one can blame the circumstances.
The external sources of the problems are not even considered.[3]
And although every problem of the system is ultimately reducible
to the difference between system and environment, this is not
even considered.

Even for the earlier theory of *societas civilis* this was no
different, and the same is true for practical philosophy: what is
social was viewed as *civitas*, as *communitas perfecta* or as political
society, even if this included all of humanity. According to the
Stoic as well as the Christian theory, non-human nature was to
be used by everyone. *Dominium terrae* thereby became a concept
by which the sacralization of all of nature was prevented and
the specification of what was religious was secured. Nature in
this sense, so ecologically important today, was de-sacralized
nature. The ever present counter-opinions were never strong
enough to present the developing natural sciences with a problem.
Then in the eighteenth century the problem experienced a
dramatic reversal. The counter-concept (which, as is so often the
case, suggests the real interest) was changed. Civilization took
the place of the sacred (whose specifically monotheistic version

was retained) as the counter-concept to nature.⁴ Thereby nature became, on one hand, an irretrievably lost history and, on the other, society's field of research.

But even this version was not enough to determine the difference between nature and civilization. It does, however, offer the first chance at an awareness of the environment (for example, as the consequence of the ancient doctrine that God is to be worshipped in his creations). The eighteenth century discovered the meaning of milieu,⁵ i.e., of being situated concretely, for example, as the connection between climate and culture. Stimulated by progress in agricultural technology, the early French economists (physiocrats) saw property as a legal institution that is both economically and ecologically ideal because it guarantees the proper treatment of natural resources while it reconciles them with human interests. It is noteworthy that at that time the *internalization* of the consequences of actions and their inclusion in rational calculation were viewed as a function of property. Today the converse is the case: the consequences of actions are discussed in terms of *externalization* and property is criticized for lacking in responsibility for these consequences.

At first nothing resulted from all this. The French Revolution led to an ideologizing of social debates linked to social position and political goals, and the descriptions of social relations occurred entirely within society. This is most clearly visible in the way in which Darwin was carried over into the social sciences. Instead of accepting the idea that the environment selectively decides how society can develop, an ideologically tainted Social Darwinism came into being that promised individuals, economies and nations the right to success through the survival of the fittest. But after a few years this too became bogged down in the mire of a new social morality, and even today the theory of evolution in the social sciences has not freed itself completely from this disaster.⁶

Even where sociology presented itself as opposition or as 'critical theory' all that it considered was society and humane principles that did not correspond to the society, or at least not at that time. This found expression as insufficient freedom, equality, justice or reason – in any event, all bourgeois themes. The part that sociology played within this social discussion was the *self-critique* of society *vis-à-vis determinate ideals*, not *frustration* regarding *uncertain hopes and fears*. But it was simply too easy to reject this critique because ideals have a fatal tendency to transform themselves into illusions. The theoretical background for this discussion has long since disappeared even though the 'simultaneity of the non-simultaneous' still had to be reckoned with for a long time. In this state of alarm the only question must be how justified are specific hopes or fears? Or, from the perspective of a disinterested observer, which factors determine the readiness to accept risks and how are they distributed in society?

Totally absorbed in its own object, sociology did not even notice that a reorientation had already started among the natural sciences, begun by the law of entropy. If this law that declares the tendency to the loss of heat and organization is valid then it becomes even more important to explain why the natural order does not seem to obey it and evolves in opposition to it. The answer lies in the capacity of thermodynamically open systems – those related to their environments through inputs and outputs – to enter into relations of exchange, i.e., environmental dependency, and nevertheless to guarantee their autonomy through structural regulation. Ludwig von Bertalanffy appropriated this idea and used it as the basis for what today is called 'general systems theory'.[7]

It would be unfair, however, to say that sociology did not take account of this at all because there are some programmatic similarities.[8] For example, research in the sociology of organizations, emphasizing the environmental reference of organizations, has been successful.[9] But here the environment always means something internal to the society, for example, markets or technological innovations, in other words only society itself.[10]

This preoccupation with society itself can be avoided only through a change in the theoretical focus of the paradigm.[11] Such a manoeuvre, however, has consequences that reach all the way into the ramification of sociological thought.[12] This means that radical incisions have to be made, and only after such an operation – if it is not refused to begin with – does one learn to proceed again slowly.

The surprising appearance of a new ecological consciousness

has left little time for theoretical consideration. Initially, therefore, the new theme was considered within the context of the old theory. Accordingly, if society endangers itself through its effects on the environment then it has to suffer the consequences. The guilty should be found out, restrained and, if necessary, opposed and punished. Moral right, then, is on the side of those who intervene against the self-destruction of society. In this way the theoretical discussion surreptitiously becomes a moral question and any of its possible theoretical shortcomings are offset by moral zeal. In other words, the intention to demonstrate good intentions determines the formulation of the problem. So, by accident, as it were, a new *environmental* ethics enters the discussion without ever analysing the all-important system structures.

Whoever proposes a new ethics, brings the question of blame into historical view at the same time. It has been advocated, for instance, that the Christian West was disposed to deal with nature in a crude and insensitive way, if not simply to exploit it,[13] while on the other hand it was also argued that Christians always loved and respected animals and paid homage to the Creator through His creation.[14] When the question is put in such a simple and naive way both of these arguments are valid. This historical perspective serves only to provide contrast and does not concern itself at all with the actual course of events. It merely helps to bring the new ethics into view without raising the difficult question of whether and how it is possible as such.[15]

Despite this, it has become obvious that as scientific research progressed respect for 'natural balances' increased, whether this was in ecological relations, foreign cultures or even today in developing countries and their traditions. But at the same time, one's own society was exposed to an incisive critique that was replete with demands for intervention, *as if it was not a system at all.*[16] Obviously this reveals a negative ethnocentrism, and it is possible that a significant aversion to 'systems theory' has had something to do with the critical restraint this theory has directed against its own society.

At the very least this summary discussion lacks an understanding of the theoretical structure of the ecological question, above all of its fundamental paradox – that it has to treat all facts in terms of unity *and* difference, i.e., in terms of the unity of the ecological interconnection and the difference of system and environment that breaks this interconnection down. As far as the ecological question is concerned, the theme becomes the unity of the difference of system and environment, not the unity of an encompassing system.[17]

Therefore the systems-theoretical difference of system and environment formulates the radical change in world-view. This is where the break with tradition is to be found, not in the question of a crude and insensitive exploitation of nature. Indeed, historical investigations of the concepts of *periechon, continens, ambiens, ambiente* and medium can show that what is today called environment was viewed by the Greek and even the medieval tradition as an encompassing body, if not as a living cosmos that assigned the proper place to everything in it.[18] These traditions had in mind the relation of a containment of little bodies within a larger one. Delimitation was not viewed as the restriction of possibilities and freedom but instead as the bestowal of form, support and protection. This view was reversed only by a theoretical turn that began in the nineteenth century when the terms '*Umwelt*' and '*environment*' were invented and which has reached its culmination today: systems define their own boundaries. They differentiate themselves and thereby constitute the environment as whatever lies outside the boundary. In this sense, then, the environment is not a system of its own, not even a unified effect. As the totality of external circumstances, it is whatever restricts the randomness of the morphogenesis of the system and exposes it to evolutionary selection. The 'unity' of the environment is nothing more than a correlate of the unity of the system since everything that is a unity for the system is defined by it as a unity.

The consequences of this interpretation for a theory of the system of society (and indeed for a system of society that communicates about ecological questions) can be reduced to two points:

1   The theory must change its direction from the *unity* of the social whole as a smaller unity within a larger one (the world) to the difference of the system of society and environment,

i.e., from unity to difference as the theoretical point of departure. More exactly, the theme of sociological investigation is not the system of society, but instead the *unity of the difference of the system of society and its environment.* In other words, the theme is the world as a whole, seen through the system reference of the system of society, i.e., with the help of distinctions by which the system of society differentiates itself from an environment.[19] After all, difference is not only a means of separating but also, and above all, a means of reflecting the system by distinguishing it.

2 The idea of system elements must be changed from substances (individuals) to self-referential operations that can be produced only within the system and with the help of a network of the same operations (autopoiesis). For social systems in general and the system of society in particular the operation of (self-referential) communication seems to be the most appropriate candidate.

If these two points are accepted then 'society' signifies the all-encompassing social system of mutually referring communications. It originates through communicative acts alone and differentiates itself from an environment of other kinds of systems through the continual reproduction of communication by communication. In this way complexity is constituted through evolution.

The considerations that follow presuppose this theory – not in order to provide a solution to the problem of the ecological adaption of the system of society, but instead to see what contours the problem takes on when it is formulated with the help of this theory.

2

Causes and Responsibilities?

Once one acknowledges the phenomenon of evolved complexity the focal point of the ecological problematic changes. The customary way of treating ecological problems begins from causes within society and then seeks responsibility for their effects. Thus it follows the normal temporal pattern and argues that the results will not occur if not preceded by the causes. Accordingly, the problems are best eliminated at their source, for example, when a chemical plant disposes of poisonous waste at a garbage dump or into a river with the consequence that fish are killed or that the water-supply becomes contaminated. An enforceable legal code suffices to handle such problems. But both the problem typology as well as a systems-theoretical analysis require a change of approach: a reconstruction of the problem from a systems perspective, one that is sensitive to the effects of ecological changes. Intercepting the causes is one of the possible ways of taking care of their effects, but only one among many. The problem of reaction to effects and the possible (almost limitless) causes and effects of such reactions remains. In other words, the 'tragedy' of decisions is that the affected system is also the cause of its own damage. But this is still not a formula for the solution of problems.[1]

Politics and jurisprudence, for instance, use the 'causer-principle' (*Verursacherprinzip*) to ascribe costs and to assign responsibility,[2] and it is clear that this produces a problem of selecting the causer. Usually, this problem is handled reflexively by appealing to the purpose of the selection itself.[3] The covert significance of the causer-principle, then, is not a causal statement but, as so often, a statement indicating a difference: alternatives (for example, subventions) are rejected because of the expense to the general public.

Science has long since left this practically significant stage of analysis behind. In the age of systems-theoretical analyses causal interconnections have been viewed as extremely complex and, in principle, opaque – unless their determination is simplified through a more or less arbitrary attribution of effects to causes. The last three decades' research into attribution shows that the real problem resides in the attribution of habits and procedures that illuminate and give importance to a selection of the many causes and effects.[4] More exactly, the determination of causes, responsibility and guilt helps to identify non-causes (*Nichtur-sachen*) and to determine innocence and the absence of responsibility too. If the producers fall on the one side then consumers must fall on the other. In this way the attribution procedure shows its real importance in providing exculpation.

All this is accepted today and does not need to be proved. The theory of self-referential systems alone, however, has realized that the classical instruments of the acquisition of knowledge, namely deduction (logic) and causality (experience), are merely forms of simplifying the observation of observations. For social systems this means forms of simplifying self-observation.[5] Methodologically, this means that the point of departure has to be the observation of self-observing systems and not the assumed ontologic of causality. In other words, one cannot avoid a decision about what counts as a cause and who is to be held responsible. It also means that morality and politics are overburdened by the unavoidability of this decision. The question then is how can this decision present itself so that the impression arises that it has not?

Radical theoretical positions of this kind lie far outside what social communication and ordinary consciousness accept today. Their consequences would require a rethinking whose results are unforeseeable. In any event, a period of slow and tedious development seems inescapable. But at the same time, the corresponding circumstances are evident – and precisely in ecological discussion itself. Attribution and assigning responsibility have consequences themselves. Political coalitions, for instance, can come apart and economic enterprises depending on them can fail. Theories and calculations which subsequently prove false can be used to justify decisions, a discovery that itself can trigger new consequences. All this may be very clear and obvious. But at present the problem is how to promote it publicly and legitimize it? Whoever observes this telescoping (*Engführung*) of observations will come to the conclusion that 'tragic' decisions of this kind have to conceal their own contingencies so that they do not have to be revealed as decisions, or at least not in certain respects.

It is well known that Walter Benjamin thought that the difference between legislation and adjudication was used for this purpose of concealment, especially in reference to his concept of violence.[6] This holds for politics as well as law. In the economic system the same function seems to be fulfilled by the difference between the determination of available quantities and the decisions of distribution under the condition of scarcity.[7] In both cases, however, it is shown that every individual decision affects the difference itself. Nevertheless, in both cases it is also true that this cannot be extolled as responsibility. The difference must be presupposed in order to determine where decisions are to be made. In effect, it replaces its own arbitrariness and resolves a paradox of self-reference. Only then can one assign responsibility.

At this point we are already in the midst of system analyses. In this chapter we have been concerned only with establishing premisses. In the analyses that follow we will ignore the question of guilt. Of course, this does not mean that its clarification from the point of view of political representability or of the legal propriety of standards is unimportant. On the level of our analyses this question would lead to the discovery that society itself is guilty – and we know this already.

3

Complexity and Evolution

'We really can change the whole thing', is a slogan that could still be heard even quite recently. Courage is all that is needed – and cybernetic guidance! Complexity has simply been exploited insufficiently until today causing all kinds of mistakes and problems for the system's output. Instead, the system has to use variety (i.e., number of possible states) to control variety and in this way, acquire the 'requisite variety for running the world'.[1] This kind of optimism seems to have passed. It underestimated the much discussed problematic of structured complexity. Above all, it did not understand that the concept of complexity itself designates a unity that acquires meaning only in reference to difference, indeed in reference to the difference of system and environment.[2]

It is not saying much to state that the world or a system is 'complex'. From this point of view everything determinate results from the reduction of complexity. Instead, one could simply say that everything occurs only in the world. But not much is gained in this way. Statements concerned with complexity become productive only when they are turned from unity to difference. The distinction of system and environment can be used to do this. It enables one to make the statement with which we will introduce the following discussion: that for any system the environment is always more complex than the system itself.[3] No system can maintain itself by means of a point-for-point correlation with its environment, i.e., can summon enough 'requisite variety' to match its environment. So each one has to reduce environmental complexity – primarily by restricting the environment itself and perceiving it in a categorically preformed way. On the other hand, the difference of system and environment is a prerequisite for the reduction of complexity because reduction can be performed only *within the system*, both for the system itself *and* its environment.

To make this matter even clearer, the question can be put as follows: how can a restrictedly complex system exist in a much more complex environment and reproduce itself? In so far as a genetic explanation is desired, the question could be handed over to the theory of evolution. Evolutionary selection can be used to explain how system structures, which maintain themselves under the pressure of complexity, develop.[4] But the theory of evolution itself has so far not provided a satisfactory explanation of this because there are obviously several, if not many, solutions to this problem and because the choice among them cannot be explained satisfactorily as environmental selection or as an adaptive performance of the system.

One possibility could be greater indifference and insulation of the system, i.e., less environmental dependence and sensitivity by restricting causal interdependencies. It is obvious, however, that macrochemical as well as organic and sociocultural evolution transcend this possibility,and it is difficult to understand how they can be made to do so by environmental complexity. Thus the question arises of what other forms can replace indifference and insulation as functionally equivalent. Again, the answer is greater system complexity.[5]

'Greater system complexity' is not a simple quality and therefore, the 'increase' cannot be understood in terms of one dimension alone.[6] This means that talk of 'more' or 'less' complexity is necessarily vague. Despite this, universally valid statements can still be formulated. Accordingly, systems with greater complexity are generally capable of entertaining more and different kinds of relations with their environments (for example, of separating inputs and outputs) and thus of reacting to an environment with greater complexity. At the same time, they have to select every individual determination internally with greater exactness. So their structures and elements become increasingly contingent. This leads to another question: which structures can meet such demands?

Evolution does not merely mean the selection, by a particular environment, of the systems that are capable of survival or increasing the adaption- and survival-capacity of systems to a particular environment.[7] This does not explain why the environment continually produces stimuli for variation and yet allows a multitude of systems to exist completely unchanged. The theory of evolution must therefore include systems theory in the explanation. Self-referentially autopoietic systems are endogenously restless and constantly reproductive. They develop structures of their own for the continuation of their autopoiesis. In this way the environment remains as the condition of their possibility and as a constraint. The system is both supported and disturbed by its environment. But it is not forced to adapt by the environment nor allowed to reproduce only through the best possible adaption. Even this is a result of evolution and at the same time a condition of further evolution.

Only when this reformulation of the theory of evolution is accepted can one use it to explain why, ecologically, the system of society is not necessarily directed toward adaptation and can even place itself in jeopardy. The system forms its own structures in reaction to irritation from the environment in order to continue the autopoietic process, or it simply ceases to exist. Thereby, it acquires the often unrealistic idea that the environment adapts to it and not vice versa. Very complex systems can develop in this way if forms of organization can be found that are compatible with greater complexity, i.e., make corresponding reduction-performances possible. The dynamics of complex autopoietic systems itself forms a recursively closed complex of operations, i.e., one that is geared toward self-reproduction and the continuation of its own autopoiesis. At the same time, the system becomes increasingly open, i.e., sensible to changing environmental conditions. All this can proceed along two lines of development:[8] the evolution of systems with greater but more reducible (more operational) complexity of their own and the increasing temporalization of autopoiesis. In the latter case autopoiesis no longer has to deal exclusively with the preservation of the existing system-state or with the continual replacement of elements that have fallen out (for example, the replication of cells or macromolecules within cells) but eventually creates systems from events whose continual passing is the necessary cause of the autopoiesis of the system.

Thus the exposure to ecological self-endangerment remains within the context of the possibilities of evolution. Threatening situations occur not only because a higher degree of specialization in answer to environmental changes reveals itself as misguided.[9] The possibility also exists that systems act on their environment in such a way that they cannot exist in this environment later on. The primary goal of autopoietic systems is the continuation of autopoiesis without any concern for the environment. Typically, the next step in the process is more important for them than the concern for the future, which indeed is unattainable if autopoiesis is not continued. Viewed from a long-term perspective, evolution is concerned about reaching 'ecological balances'. But this merely means that systems pursuing a trend toward exposure to ecological self-endangerment are eliminated.

If this evaluation of the evolution of social complexity and ecological problems is correct, then the question of the 'domination of nature' has to be reformulated. It is no longer an issue of a greater or lesser technological control over nature or even of sacred or ethical road-blocks. Nor is it a matter of the protection of nature or of a new taboo. To the extent that technological intervention changes nature and problems result from this for society, *greater* rather than *less competence for intervention* has to be developed, but practiced according to criteria which *include reaction on itself.* The problem does not lie in causality but in the criteria for selection. The question that follows from this is twofold: (1) is there enough technological competence for selective behavior, i.e., does it give us enough freedom *vis-à-vis* nature? (2) is there enough social, i.e., communicative, competence to be able to carry out the selection operatively?

4

Resonance

Concepts like complexity, reduction, self-reference, autopoiesis and recursively closed reproduction with environmentally open irritability raise complicated theoretical questions that cannot be pursued in all their ramifications in what follows. So we will simplify the presentation by describing the relation between system and environment with the concept of *resonance*. We will also assume that modern society is a system with such a high degree of complexity that it is impossible to describe it like a factory, i.e., in terms of the transformation of inputs into outputs. Instead, the interconnection of system and environment is produced through the closing-off of the system's self-reproduction from the environment by means of internally circular structures. Only in exceptional cases (i.e., on different levels of reality, irritated by environmental factors), can it start reverberating, can it be set in motion. This is the case we designate as resonance. One can imagine a dictionary that would define nearly all the concepts that it uses by referring to other definitions and would allow reference to undefined concepts only in exceptional cases. An editorial committee could then be formed which would supervise whether language changes the meaning of those undefined concepts or, through the formation of new ones, disturbs the closure of the lexical universe without determining how changes in the entries are to be handled when this disturbance occurs. The richer the dictionary, the more it is kept going by the development of language, i.e., the more resonance it will be able to produce.

Physics can also be called on to help us. A differentiated system can be made to resonate only on the basis of its own frequencies. In the biological theory of living systems, 'coupling' is used to indicate that there never are point-for-point correlations between the system and environment. Instead, the system uses its boundaries to screen itself off from environmental influences and produces only very selective interconnections.[1] If this selectivity of resonance and coupling did not exist the system would not be able to distinguish itself from the environment. It would not exist as a system.

The same is true for the process of communication in the social system. We can formulate the question of the ecological basis of and danger to social life much more exactly if we look for the conditions under which the states and changes in the social environment *find resonance* within society. This is by no means something that is more or less self-evident. On the contrary, it is improbable according to systems theory. From the evolutionary point of view one can even say that sociocultural evolution is based on the premiss that *society does not have to react to its environment* and that it would not have taken us where it has if it proceeded differently. Agriculture begins with the destruction of everything that had grown there before.

We find the problematic of a purely selective contact with the environment and the use of boundaries for screening-off on the level of individual system operations too. Society is a system, *sit venio verba*, uncommonly rich in frequencies. Everything that can be formulated linguistically can be communicated about. But we remain bound to language (just like we are bound to the narrow spectrum of what we can see and hear), and what is more important and decisive, speech and writing have to be ordered sequentially. Everything cannot be said all at once nor can all statements be connected with all others. The general structure of language (its vocabulary, grammar and the way it uses negation) makes selections necessary. This means that all selections themselves have to be ordered sequentially, i.e., appear in a context of succession in which one phrase makes another intelligible but never the whole. Even if no boundaries of the social system were given, i.e., even if we could start from the very beginning with society, communication's mode of operation would establish boundaries simply through its coming into being and continuation. If communication takes place then this *eo ipso* differentiates a social system whether anyone wants it and approves of it or not.

These constraints on the social system's capacity for resonance are attuned to the mode of information processing that society and psychical systems apply in common: to the characteristics of meaning. The possibilities of a meaningful grasp of the world are themselves attuned to – and then require – the necessity of a purely momentary grasp of the world at any time. Only very little can form the actual focus of attention or be treated as an actual theme of communication. Everything else, including the world as a whole, is associated with this only by means of references, i.e., accessible only sequentially and selectively. Only one of these possibilities can be pursued at any time, and every advance creates more possibilities than can be handled subsequently.[2] This is what Husserl meant when he described the world as the 'horizon' of actual intentions. It is actual as a horizon, never as a *universitas rerum*. One can see in this a formula, as it were, for the insolubility of ecological problems, even though, at the same time, it is known that every reference leads to something determinate or determinable[3] – that there are no paradoxes.

In other words, meaning is a representation of world complexity that is actualizable at any moment. The discrepancy between the complexity of the actual world and consciousness's capacity for apprehension or communication can be bridged only when the scope of the actual intention is restricted and all else is rendered potential, i.e., reduced to the status of mere possibility. There is no such thing as a 'stimulus inundation' since the neurophysiological apparatus already screens off consciousness drastically, and the operative medium of meaning has to work very hard to permit something that is well digested to become actual. So the established view of anthropology has to be revised. We will put the idea of the very restricted resonance capacity of meaningfully constituted, operatively closed systems in its place.

In the case of meaning-processing as well as living systems[4] autopoiesis has to be secured before all else. This means that the system exists only if, and as long as, meaningful information processing is continued. We can designate the structural technique that makes this possible as a *difference technique*.[5] The system introduces *its own distinctions* and, with their help, grasps the states and events that appear to it as *information*. Information is thus a purely system-internal quality. There is no transference of information from the environment into the system. The environment remains what it is. At best, it contains data. Only systems can *see* the environment because this requires the seeing of other possibilities, the presence of a pattern of difference and the situating of items within this pattern as a 'this instead of that'. In the environment there is no 'instead of that', no 'this' as a selection out of other possibilities, i.e., neither a pattern of difference nor information. To emphasize this once again:[6] system boundaries have to be drawn so that the world acquires the possibility of observing itself. Otherwise there would be pure facticity alone.

In a somewhat different terminology one could say that system differentiation makes possible the establishment and reduction of complexity. The system can place possibilities within the environment and view what is found there as a selection from numerous possibilities. It can project something negative and use this to identify something positive. It can form expectations and be surprised. All these are structures for the operation of systems themselves. They presuppose that the system can distinguish itself from the environment.

If physical systems have differentiation and highly selective resonance at their disposal then this is certainly the case for meaning-constituting systems too, especially society. The difference technique can be used by these systems because distinctions, negations, possibility projections and information are and remain purely internal and because, in this respect, no environmental contact is possible. In this way the systems remain dependent on autopoiesis, on a continual self-renewal of their elements by their elements, but because information and information expectations, i.e., structures, are obtained by means of difference projections, this closure is openness at the same time. For the system can experience itself as its difference from the environment by means of the very same difference technique.[7] This in no way changes the internal closure of the interconnection of its own operations. Instead, this equips it with the capacity to react to whatever is environment *for it.*

This theoretical account brings us to the following question: which concepts and distinctions in social communication help us to deal with the exposure to ecological dangers? It excludes the very obvious and ordinary idea that there are facts that call for reaction or else damage will result. But even facts have communicative effect only as facts, and the establishment of a fact is the establishment of a difference.[8] Therefore we have to ask in which difference patterns are facts grasped, which desired states bring states into relief and how do expectations become accustomed to whatever appears as reality to them?

In addition to this so-called 'constructivist' perspective, the social system's differentiation must be kept in mind. It is just as suggestive as it is misleading to assume that 'the' system reacts to 'the' environment, even if this is only to 'its' own idea of 'the' environment. The system/environment difference is indeed the presupposition of all observation of the environment. But this does not mean that the system as a closed unity can react to the environment. The unity of the system is nothing more than the closure of its autopoietic mode of operation. The operations themselves are necessarily individual operations within the systems, i.e., several among many others. There are no all-encompassing operations. Besides, complex systems like societies are differentiated into subsystems that treat other social domains as their (socially internal) environment, i.e., differentiate themselves within the society, for example, as a legally ordered political system that can treat the economy, science, etc. as environment and thereby relieve itself of direct political responsibility for their operations.

This differentiation theorem has far-reaching consequences. It implies:

1 That important performances of the societal system are constantly executed by *subsystems* because this is the only way to achieve a sufficient level of complexity, and that in order to explore how a society can react to the exposure to ecological dangers the constraints on the possibilities of its subsystems must be examined. These in turn depend on the form of social differentiation.

2 The system's unity can, if necessary, be *represented* within the
   system itself, where the concept of representation is understood
   as a *representatio identitatis,* and not as a taking-the-place-
   of-something-else. Representation is the reintroduction of the
   system's unity within the system itself. This creates a *difference*
   within it, whether this is sought or not.[9] The presentation of
   the system's unity within the system itself must therefore fit
   the pattern of system differentiation. It may appear as the
   'top' if the system is differentiated in a hierarchical manner,
   i.e., presents itself as stratification. Or it may appear as the
   'center' if the system is differentiated according to the center/
   periphery pattern (for example, city/countryside). It cannot
   choose any of these forms of presentation if none of these
   forms of differentiation exists. We will also have to consider
   whether there are further possibilities and whether the exposure
   to ecological danger could be an occasion to develop other
   possibilities.

3 Since every operation is only one among many, every
   operation within the system is observable by others. Formally,
   observation means being treated as information on the basis
   of a pattern of difference, normally through expectations that
   are fulfilled or not. In this sense, self-observation constantly
   accompanies the operations taking place within society. This
   observation creates additional effects of its own, often in
   opposition to those that the operations themselves intend.
   Thus, on the one hand, there can be an immediate stifling of
   initiated plans and, on the other, an effect-explosion which
   neither waits for nor depends on the operations reaching their
   intended goals.

   Further inspection reveals that the theory of self-referentially
closed and thereby open social systems leads to considerable
complications. The concepts of system differentiation, representation and self-observation indicate what in particular needs to
be clarified to understand whether and how society can create
resonance because of exposure to ecological dangers. But it is
already clear that the problems cannot be solved merely with
admonitions and appeals to more environmental consciousness.

Instead, the observation accompanying all political, economic and scientific operations, and precisely from these perspectives, may trigger one of those 'effect-explosions' that change society – entirely independently of whether the relation of the social system to its environment is improved in this way and, if it is, according to what criteria.

5

The Observation of Observation

System resonance, then, is always in effect when the system is stimulated by its environment. The stimulation can be registered by the system if it possesses a corresponding capacity for information processing permitting it to infer the presence of an environment. Similarly, the system registers the effects of its own behavior on the environment whenever this behavior triggers a stimulation within the range of the system's possible perceptions. The environment is the total horizon of information processing that refers beyond the system. Thus it is an internal premise for the system's own operations constituted within the system when the latter uses the difference of self-reference and other-reference (or 'internal' and 'external') to order its own operations.

As an internal premise, the system's environment has no boundaries nor needs any. Presenting itself as a horizon, it is the system-internal correlate of all references that extend beyond the system. This means that, whenever necessary, any operation can push it back still further. The horizon always recedes when it is approached, but only in accordance with the system's own operations. It can never be pushed through or transcended because it is not a boundary. It accompanies every system operation when this refers to something outside the system. As a horizon, it is the possible object of intentions and communication; but only in so far as the system can present the environment to itself as a unity – and this requires that it can differentiate itself as a unity from it.

In a somewhat different, Wittgensteinian formulation one could say that a system can see only what it can see. It cannot see what it cannot. Moreover, it cannot see that it cannot see this. For the system this is something concealed 'behind' the horizon that, for it, has no 'behind'. What has been called the 'cognized model'[1] is the absolute reality for the system. It has a singular quality of being or, logically speaking, univocality (*Einwertigkeit*). It is what it is, and if it turns out that it is not what it seems to be then the system has made a mistake! The system can operate only with two values when it uses the distinction of self-reference and other-reference.

All this necessarily holds true for a system's immediate observation of what presents itself as environment to it. Nevertheless, a system that observes *other* systems has other possibilities. Even if it posits its environment apodictically, like every other system, the observation of a system by another system – following Humberto Maturana we will call this 'second-order observation'[2] – can also observe the restrictions forced on the observed system by its own mode of operation. The observing system can discover that the environment of the observed system is not constituted by *boundaries* at all, but, perhaps, by *constraints*. It can observe the horizons of the observed system so that what they exclude becomes evident. Using this, it can clarify the mode of operation of the system/environment-relations in a kind of 'second-order cybernetics'.[3]

At present, second-order cybernetics seems to be the place where the problems of the foundations of logic and epistemology can, at least, be handled if not 'solved'. We will therefore have to examine it briefly because these problems become increasingly important when we deal with science as a part of society, i.e., as part of the object discussed by it (including these words as part of this text too).

Since social systems in general and societies in particular constitute themselves through autopoietic self-reference, every observer is confronted with the question of how these systems come to terms with the problems of tautology and paradox that necessarily follow when a system operates through self-reference *alone*, i.e., when it must ground *all* its operations in self-reference. The classical answer to this problem (Russell and Whitehead, Tarski) is well known. Such a system has to interrupt or 'unfold' self-reference while it distinguishes several levels as a hierarchy of types, i.e., while it separates object-language, meta-language and, if necessary, meta-meta-language. This solution does not work, however, because the concept of level assumes a plurality, i.e., a reference to other levels. This means that operations capable of interrupting the hierarchy through the performance of a 'strange loop' cannot be eliminated.[4] The hierarchy of levels can be saved only by an arbitrary fiat: the instruction to ignore operations that disobey the command to avoid paradoxes. Questions 'why?' are not permitted despite the constant temptation to raise them.

Since this requires the rejection of universalistic theories (and would put us in the embarrassing position of never being able to discover that society can summon only limited resonance to its environment) another solution has to be found. The fatality of the arbitrary proscription of thematizations aimed at avoiding paradox can itself be avoided only through the distinction of natural and artificial constraints on self-reference.[5] Those contraints on a system's self-reference that appear as natural or necessary are the ones that are the conditions of the possibility of operations, i.e., that conceal tautology or paradox in the performance of self-reference. Those for which this is not the case are artificial or contingent.

This distinction is always to be treated as system-relative although it can be viewed as variable too. The constraints that previously seemed necessary become artificial when learning processes reveal how they can be replaced as eliminators of tautology and paradox. In this context second-order cybernetics is important.

An observer who recognizes that an object is a self-referential system notices at the same time that it is constituted tautologously and paradoxically, i.e., is arbitrary and inoperable, unobservable. This produces a paradox of its own for the observer of a self-referential system: the arbitrariness and impossibility of observation. The observer can avoid this embarrassment by distinguishing natural and artificial constraints applied to the system observed. Then it becomes clear that this system cannot see what it cannot see. For the observer whatever is necessary and irreplaceable in the system can appear as contingent. With the assumption of let us say, a supermodal, observationally dictated distinction of necessary and contingent the observer can eliminate the paradox while providing the observation with an operational object. This can be done in a way that implies learning possibilities for the object – through the possibility of the displacement of the boundary line between natural and artificial constraints of complete self-reference.

Our concept of resonance assumes this second-order cybernetics. It implies constraints, presupposes a reality that triggers no resonance at all within the system and nevertheless presents an environment for this second-order observation. From this perspective one can see that the observed system constructs the reality of its world through a recursive calculation of its calculations,[6] and since this is the case on the level of living, neurophysiological and conscious systems it cannot be different for social systems either. Second-order cybernetics can be used to prove this. Consequently, it can draw no other conclusion than that this applies to its own observation too, but at the same time it can still see that what cannot be seen cannot be seen.[7]

Sociopsychological investigations of attribution have come to similar conclusions entirely independent of this biologicocybernetic research tradition. Here, research proceeds under the title of causal attribution. The actor's mode of attribution (first-order observation) is distinguished from that of the observer's (second-order observation). While the actor finds the bases for action primarily in the situation itself, the observer sees the actor-in-the-situation, looks for differences in the interpretation of the situation by different actors and makes attributions primarily in terms of the personal characteristics of the actor.[8] Correspondingly, sociology has always concerned itself with actors who already knew why they acted and, therefore, had to justify an additional, transcendent, 'critical', knowledge-guiding interest (*Erkenntnisinteresse*).[9] In all these cases the beginning has to be – and this constitutes the innovation *vis-à-vis* the naive faith in science – the fact that second-order observation together with its theoretical apparatus is possible only as a performance of structured autopoiesis, i.e., it is not 'objectively better' knowledge but only a different knowledge that takes itself for better.

To analyse the problem of the exposure to ecological dangers with the necessary exactness, second-order cybernetics must be taken as the starting-point. If the starting-point were an 'objectively' given reality that, for the time being, was still full of surprises and unknown qualities then the only issue would be to improve science so that it could know the reality better. But then the relations of the other systems to their environment – for even within society there are many other systems – would not be grasped sufficiently. Even science would not be able to understand why with its 'better knowledge' it often finds no resonance within society because what it comes to know – its 'better' knowledge – would have no value at all as reality in the environment of other systems or is at best a scientific theory for them.

Not much is gained, therefore, by following an ontological theory of reality (which corresponds to a first-order observation of the environment) because this theory is not in a position to grasp the problem as such. We have to choose a second-order cybernetics as the point of departure. We have to see that what cannot be seen cannot be seen. Only then can we discover why it is so difficult for our society to react to the exposure to ecological dangers despite, and even because of, its numerous function systems.

To the extent that society can differentiate structurally an observing of observing and explain this theoretically it finds itself in the position of establishing the conditions under which it will react through its respective (function) systems to whatever is environment for them. This is not a question of creating a basis for better possibilities of action. Nor is it even something like a 'domination-free discourse'. Inasmuch as this idea is an improvement over the old, unsophisticated way of watering the tree of freedom with the blood of tyrants, the real problem is to be found neither in a lack of justifications nor in the pattern of coercion and freedom. Nor is it even to be found in removing barriers to rational consensus and harmonious coexistence. The problem is the acquisition of a different kind of insight.

In many ways modern society has opened up possibilities for observing and describing how its systems operate and under what conditions they observe their environment. The only drawback is that this observing of observing is not disciplined enough by self-observation. It appears as better knowledge. But in reality it is only a particular kind of observing of its own environment.[10] Under these conditions the idea that rational consensus ought to be attained is quickly trivialized. Those who think they know that this is going to be a protracted enterprise use this idea and test their willingness to make concessions according to their own judgement. But every operation and every observation has structural limitations, which is precisely what second-order observation makes clear. A better evaluation of the situation is attainable only when this insight is applied to itself, i.e., is employed recursively. When this is done the constraints on the ability to observe, describe and turn insights into operations have to be analysed and compared. Any protest against such constraints would be strangely naive and, as such, would merit observation itself – if not by the protester then at least by others who observe the protester.

6

Communication as a Social Operation

In the following I will not consider the very limited, socially dependent possibilities of the *consciousness* of individual psychical systems and will make society the sole system reference. By society I mean the most encompassing system of meaningful communication. Any limitation, for example, to organizations,[1] would restrict the investigation too much. So the question is how, as an operatively closed system of meaningful communication, does society communicate about its environment? More specifically, what are its possibilities for communicating about exposure to ecological dangers?

We must be careful in our presentation of the concept of exposure to ecological danger as long as we do not know what it is about. So we will understand it very broadly. We will take it to designate *any communication about the environment* that seeks to bring about a *change in the structures* of the *communicative system that is society*. It should be noted that this is a phenomenon that is exclusively internal to society. It is not a matter of blatantly objective facts, for example, that oil–supplies are decreasing, that the temperature of rivers is increasing, that forests are being defoliated or that the skies and the seas are being polluted. All this may or may not be the case. But as physical, chemical or biological facts they create no social resonance as long as they are not the subject of communication. Fish or humans may die because swimming in the seas and rivers has become unhealthy. The oil-pumps may run dry and the average climatic temperatures may rise or fall. As long as this is not the subject of communication it has no social effect. Society is an environmentally sensitive (open) but operatively closed system. Its sole mode of observation is communication. It is limited to communicating meaningfully and regulating this communication through communication. Thus it can *only expose itself to danger*.

To formulate this important starting-point differently, one could say that the environment of the social system cannot communicate with society. Communication is an exclusively social operation. On the level of this exclusively social mode of operation there is neither input nor output. The environment can make itself noticed only by means of communicative irritations or disturbances, and then these have to react to themselves. Just as one's own lived-body cannot announce itself to consciousness through conscious channels but only through irritations, feelings of pressure, annoyance, pain, etc., that is, only in a way that can produce resonance for consciousness. Using concepts introduced by Francisco Varela[2] one can say that there is no coupling by input, only coupling by closure.

This is also true – and this gives the argument a particularly incisive importance – for the *relation of consciousness and communication*. Even the consciousness of psychical systems belongs to the environment of the societal system. As such, it is only a psychical, not a social fact. Of course, human consciousness and human life, like so much else, belong to the indispensable conditions of social communication. But this does not change the fact that, as the production of ideas by ideas, the processes of consciousness *are not* communications.[3] (Husserl saw a proof of the transcendentality of consciousness in this discovery; we merely infer from it a different system reference.) Thus, once again, the relation of conscious systems and social system has to reckon with a resonance threshold that selects in a very rigorous way. Whatever 'ecological awareness' may occur empirically within a consciousness, it is still a long way from this to a socially effective communication. Afterwards, this difference between communication and consciousness can itself become a theme of communication. But then the communication is about 'alienation', 'apathy', the resignation or protest of youth or similar themes connected only indirectly with the exposure to ecological dangers.

Viewed realistically, we have to reverse the idea that a 'subject' must first consciously resolve to communicate in order to act communicatively.[4] Only when (and for reasons that cannot be attributed to a consciousness) ecological communication is set in motion and begins to co-determine the autopoiesis of social communication can one expect the themes of this communication gradually to become conscious contents too. This simply means that social communication changes its environment, in this case mental states. What follows from this for society can be grasped only through an analysis of possible communication, through an analysis of the social system's capacity for resonance.

Conscious systems, therefore, are limited to producing irritations, disturbances or evasive themes if they have not first adjusted themselves to the social conditions of communicability. In this case the sharp delimitation of what is communicable socially signifies comprehensibility or noise. In triggering social communication-processes consciousness is guided either by the corresponding valid structures (including the structurally given possibilities of changing the structures) or it merely creates noise which, according to the possibilities of social communication, is eliminated or transformed into something communicable. This statement should not be misread as the assumption of a static system. On the contrary, a communication system's structures are highly flexible. They can be varied in their use and can even be used in a counter-sensical way, for example, ironically or for guiding deviant behavior. All this changes nothing about the fact that the threshold of possible and possibly understandable or even possibly successful communication works in a highly selective way, i.e., rejects whatever cannot find resonance.

Not least important is the realization that conscious systems, which are outside the domain of linguistic (thus communicable) articulation and depend on perception and intuitive presentation, are hardly in a position to order trains of thought into temporalized complexity. Even if an 'ecological awareness' would arise in some conscious systems or others it would have properties that would be almost useless for society. It would be overdetermined perceptually or intuitively. In any event, this is what the underlying systems-theory concludes. It would be more likely to present its ecological theme negatively through particular proposals than to present it in communication as positive knowledge about the environment. It would either tend to anxiety and protest or to a critique of society that would be unable to deal with its environment adequately. It would reach its generalizations only in a negative way and lean toward an emotional self-certainty typical of cases in which it no longer knows. In addition, it would depend on socially given forms and connection capabilities (*Anschlussfähigkeiten*) or persist in an ever-possible negation that produces little of value.

7

Ecological Knowledge and Social Communication

As surprising as it may sound we still adhere provisionally to the idea that society can expose only itself to ecological danger. This means not only that it changes the environment in a way affecting the continuation of social reproduction on the contemporary evolutionary level but also if we disregard the unlikely case of a radical extinction of all human life, that it can endanger communication only through communication. Consequently, somehow and somewhere society will have to thematize the connections between its own operations and environmental changes as problems of continued operations, if only for the sake of finding resonance within social communication. So the key question becomes how society structures its capacity for processing environmental information.

As far as I know this question has been raised and discussed only for relatively simple societies, those living on an archaic level.[1] These societies were able to present supernatural matters better than natural ones and therefore they sought ecological self-regulation in mythico–magical ideas, i.e., in taboos and rituals dealing with the environmental conditions of survival. The famous pig cycle – not the one of the political economists but that of the Maring living in New Guinea[2] – is a classical example of this. Whenever the pig population increases too much and they begin to ravage the food-supply strong ritualized justifications come into play to arrange a great feast that re-establishes a balance in the number of pigs and regulates the protein consumption of the tribe. A surprisingly pragmatic attitude toward sacred matters makes it possible to balance the environmental conditions of the system without making this requirement explicit. Among other things this means that other, functionally equivalent solutions of the problem of periodic imbalances – for example, when population increases, to cultivate better-protected gardens and to increase the pig population at the same time – are not even considered. A ritually regulated society's own structures do not program it toward growth.

Of course, these archaic societies possessed the necessary knowledge for survival and productive know-how. They certainly knew that pigs ravaged gardens, and they also knew that excessive use of the land decreases the harvest and makes the land unusable. But the semantic organization of this knowledge and its connection with the motivational guidance of human behavior was left to a semantics of the sacred – precisely because supernatural matters are easier to organize than natural ones. In this way uncertainties and prescribed social reactions could be intercepted and transformed into social certainties, and one could deal more or less successfully with the circumstance that reactions to environmental problems within society have diverse effects, i.e., favor or frustrate one more than another.

Archaic social systems were also responsible for important and irreversible environmental changes. The desolation brought about by deforestation is a good example of this, and demonstrates that the problem is not new. The extent of the possibilities as well as the social pressure to exploit them, however, have grown enormously. Besides, and perhaps even more significantly, the latent premises of a religious guidance of society have evaporated in the transition to modernity, i.e., the premises functioned only with the help of mystifications. Religious semantics always had to operate[3] with secrecy or with strategically placed vagaries. Ignorance and its accompanying uncertainty were exposed to a process of semantic atrophy. Thus it contracted into a small residue of vague indeterminacy (for example, Divine Will) and acquired a form that could be ignored.

It is evident that modern society no longer treats ecological problems in this way, no longer solves or at least modifies them via a semantics of the sacred. The transformations of cultural and religious semantics that occurred as a result of writing, alphabetizing and printing hardly permit dealing with environmental problems via taboos and ritualization any longer.[4]

Powerful counter-movements tried to do this and at the very beginning of modernity sought ultimate answers in the arcane. Erasmus, for example, pleaded against Luther for the self-confinement of religious texts in the interests of human freedom.[5] Hermetics employed ancient tradition and staked everything on what was concealed in secrets.[6] But this was all in vain! Printing and widespread dissemination confer a completely new form on technological information, formulas and the explanation and handing-down of directions. This means that knowledge now has to be understandable in itself. It has to present itself as differentiated and thereby increasingly exposes itself to comparison and correction.[7] Reference to ancient secrets, distant authorities or awe-inspiring mysteries are confronted with wanting-to-know-more-precisely. The idea of the cosmos, inherited from the Platonic tradition, as a large, visible and yet unfathomable organism, collapsed. Because of writing's duplication of oral language and the changing demands on important communication it is no longer possible to reconcile knowledge with motivation through mysteries and secrets. The traditional figures evoking respect and awe are no longer effective and cannot take the place of a knowledge that is exact and proven.[8] Mysteries are reserved for Holy Scripture. In any event, their use in everyday life shows a contempt for the communication partner.[9]

Even if these hypotheses are correct, namely that writing, the alphabet and printing actually stimulated profound changes in the communication system of society they still do not supply an adequate description of the contemporary situation and its chances for dealing with ecological problems communicatively. Theoretical tools that are more complex are necessary to describe modern society. New dissemination techniques for communication are an essential, but only one factor among these. Another is the change in the primary form of society's differentiation from the stratification of lineage, clans and families to the differentiation of function systems. This means that, today, each of the most important subsystems of society is directed to a specific and primary function that pertains to it alone. This formative principle explains the enormous growth of modern society's performance and complexity. At the same time it reveals the problems of integration, i.e., of the negligible resonance capacity among the subsystems of society as well as the relation of society to its environment. The theory of functional system-differentiation is a far-reaching, elegant and economical instrument for explaining the positive and negative aspects of modern society.[10] Whether it is correct is an entirely different question.

8

Binary Coding

We can now formulate our question more exactly. How can environmental problems find resonance in social communication if society is differentiated into function systems and can react to events and changes in the environment only through these? After all, in such a system there is communication that is not coordinated functionally or coordinated only ambiguously – the communication of the streets, so to say, or in somewhat more high-sounding jargon: 'life-world' communication.[1] Communication that affects society, however, depends on the possibilities of the function systems. We will therefore have to investigate these first because this is the only way to consider realistically what possibilities exist for communication in a society that distances itself consciously from all function systems – either through protesting, moralizing or through a blurring of differentiation.

The most important function systems structure their communication through a binary or dual-valued code that, from the viewpoint of its specific function, claims universal validity and excludes further possibilities. The classical example of this is the binary code of logic used by science. Analogously, the legal system operates with a code of legal and illegal. The economy uses property and money to distinguish clearly between possession and non-possession so that long-term possibilities of the transfer of commodities and money can be organized and calculated, and politics is guided by the questions of power that accompany governmental authority and which are put to the vote using ideological codes like conservative versus progressive or restrictive versus expansive.[2] The significance of these functional domains for the modernization of contemporary society will become evident immediately once we approach the problem of steering communication through binary codes.

As one can see, from the standpoint of a second-order cybernetics, i.e., from the observing of observations, every binary code resolves tautologies and paradoxes for the system that operates with this code. The *unity* that would be unbearable in the form of a tautology (for example, legal is legal) or in the form of a paradox (one cannot legally maintain that one is legal) is replaced by a *difference* (in this example the difference of legal and illegal). Then the system can use this difference to steer its operations. It can oscillate within it, and develop programs that regulate the coordination of the operations to the positions and counter-positions of the code *without ever raising the question of the unity of the code itself*. When this is achieved self-reference can unfold itself and does not have to be enlisted immediately and directly as unity (although within the code it comes into play dialectically, as it were, since every position is identified in reference to its counter-position). At the same time one should remember that an observer – a position within which we presently find ourselves – can see through this entire manouevre. Nevertheless, the possibility of observation arises for the observer only because a system (or a hierarchy or other functionally equivalent solutions of the problem) chooses a code to conceal those aspects of its self-reference that would reveal the tautology and paradox of its operational bases.

Binary codes are duplication rules. They form within the communication process when information acquires value and is exposed to a corresponding counter-value. The reality that is treated according to the code is singular. But it is, as it were, duplicated fictively so that every value can find its complement and be reflected in it. Of course, there are no negative facts. The world, after all, is what it is. But by coding communication about reality one can treat everything that becomes a subject of communication contingently and reflect it in a counter-value. This complementarity is not a matter of increasing or decreasing the number of facts, of an 'Another beer, please!', but of projecting a positive/negative distinction with whose help the possibility and consequences of the opposite can be examined. Thus it is neither a question of an exclusively communicative accomplishment nor of a state of affairs in the world that needs to be depicted in the communication.

Binary codes of this kind can be viewed as highly successful and important evolutionary achievements that have only attained their contemporary degree of abstraction and technical proficiency after a long development.[3] At least the most important characteristics of this structure deserve to be mentioned.

1 Codes are *totalizing constructions*,[4] i.e., *all-encompassing constructions* having a claim to universality and possessing no ontological limit. Everything that falls within their domain of relevance is assigned to the one value or the other, *tertium non datur*. Just as God Himself stood outside the Creation in creating the difference of heaven and earth so a third possibility exists for codes – at best only parasitically (parasitically is meant here in the sense in which Michel Serres uses the term).[5]

2 As the reference to everything that can be treated as information within the code, totalization leads to the *contingency of all phenomena without exception*. Everything appears in the light of the possibility of its counter value, as neither necessary nor impossible. Any necessities or impossibilities have to be reintroduced in a counter-move – perhaps to remove paradox from the code (see 4 below) – and therefore remain doubtful.

3 Codes are *in so far-as-abstractions*. They are valid only in so far as communication chooses their domain of application (which, by the way, it does not have to do). After all, not every situation is a matter of truth or justice or property. Thus the use of a code is a socially contingent phenomenon since this is the only way it can totalize a schema that reduces everything to two opposed possibilities.[6]

This produces a connection between coding and functional specification in the process of evolution: certain binary codes are used only when the operations to be coded occur in the corresponding function-system. Just as, on the other hand, social function-systems attain universal relevance for all the operations concerning them because they are specialized according to the operations of a determinate code.

4 Codes, as mentioned already, *resolve the paradox* inherent in the problematic underlying every self-referential relation. Yet every coding leads to the problem of applying the code to itself and thereby, sometimes, to a paradox. Logical antinomies like the liar's paradox – 'This sentence is a lie' – are well known. But other codes have similar problems too: for example, by what right is the difference of legal and illegal introduced and upheld? Another example of paradox can be found in the increasing dependence of those occupying a greater position of power on help from others. In the parliamentary code of ruling versus opposition parties the ruling party will often show an inclination to undercut the opposition by anticipating its position and taking it, or capital finds itself under the constant pressure to reinvest, i.e., under the constant pressure to facilitate the consumption of others. When the code is transformed into a contradiction it deactivates this problem for the operations that the code regulates. 'A because non-A' becomes 'A is non-A' and in this form the problem is eliminated. Such contradictions – allowing for some residual problems that can be left aside for special treatment – are avoidable. But precisely in this regard codes remain sensitive to changes in the conditions of social plausibility.[7]

5 Coding uses and reinforces the old adage that *opposites attract*,[8] or in the words of latin rhetoric, *Contrariorum est eadem disciplina*. Difference integrates. The transition from one side to the other is pre-programmed in the difference-scheme and thereby facilitated. Negation is all that is needed to accomplish this for the logical code. The operative proximity of value and counter-value leads almost necessarily to the differentiation of corresponding function-systems. It is simple to transform property into non-property through exchange or sale. It is much more difficult, however, to submit it to a legal examination or apply it politically.

6 In binary coding the guiding value of the code (truth, justice, property, etc.), has at the same time to surrender its right to

serve as the *criterion of selection* because to do so would contradict the formal equivalence of position and negation. After all, the establishment of falsity can have a much more positive effect upon the advancement of science than the establishment of truth. Very often this is simply a matter of the theoretical context.[9] Property can become a burden when it is just an expense and not a source of revenue. This is determined completely by the context of investment, and even the government is inclined to renounce responsibility for certain political decisions. This again is determined entirely by the kind of political programs (policies). Criteria, therefore, cannot be frozen into the abstractness of the code because they are not designed to establish the possibility of functionally specified operations. Instead, they serve much more concretely to steer correct and useful operations. The code, therefore, can outlast the change of criteria (and, in principle, of all criteria) although it is hard to imagine that everything could be changed all at once and that the code could be suspended for the purposes of a completely new beginning.

7 The difference of code and criteria for correct operations (or of coding and programming) makes possible the combination of *closure and openness in the same system*. In reference to its code, the system operates as a closed system; every value like 'true' and 'false' refers to its respective counter-value alone and never to other, external values. But at the same time, the programming of the system makes it possible to bring external data to light, i.e., to fix the conditions under which one or the other value is posited. The more abstract and technical the coding, the richer the multiplicity of the (internal) operations with which the system can operate as closed and open at the same time, i.e., to react to internal and external conditions. One can also designate this as an increase in resonance capacity. But no matter how 'responsive' the system may be structurally[10] and no matter how sensitive its own frequencies, its capacity for reaction rests on the closed polarity of its code and is sharply limited by this.

8 Coding effectively excludes third values although these may be reintroduced into the system on the level of programming

correct behavior – of course, only under the conditions valid for this level. Despite the explosiveness of new themes, a threefold code, perhaps of the type true/false/environment or legal/illegal/suffering, is never a possibility. Nevertheless, environmental problems can become the object of research programs or human suffering and their prevention the object of legal regulations. Thus the differentiation of coding and programming makes the reappearance of the third value possible; but only to co-steer the allocation of the code-values on which it primarily depends.

9 Furthermore, coding signifies the *bifurcation* of operations and the structures built on them with the well-known consequence of the constitution of *historically irreversible complexity*. The self-reflecting distinction established within the code produces consequences that are based on the fact that truth is not falsity and that property or political power can be transformed into their opposites only through determinate procedures like exchange or voting. The structured complexity that comes into being in this way is not under the system's control. Neither can it be grasped as a unity nor can the code be applied to itself. In other words, within the system one cannot decide whether all falsities are false, whether all injustice is illegal and the expropriation of property is conceivable only as revolution (and then realizable only as a partial transferral within the economic system).

10 Coding channels all further information processing into its domain and is guided primarily by its initial distinction because this is the only way to generate information and co-ordinate it with a function system. All further information processing transforms differences into differences,[11] for example, in determining whether a particular investment of a capital sum will be profitable and whether a corresponding demand could be generated in the market so that a price can be fixed.

11 When coding possesses all these characteristics it is technically the most effective and successful form of *differentiating function-systems*. But this does not mean that such a univocal coding is the only way to form function systems. The education

system, for example, uses a rather unwelcome code to meet the demands of its selections and has an entirely different basis in the school's complexes of organization and interaction. Similar questions could be addressed to the coding of religion. But this does not mean that, historically, the code is established first and then a corresponding system is formed. Evolution creates its own conditions as it progresses and comes to a halt when and as long as this does not succeed. For a description of modern society one will have to admit that important and distinctively modern function systems have become identified through a binary code that is specifically valid for each of them. They know, in any event, what their guiding difference is and how it functions in the operations of the system. Whether they can show what the meaning of their unity is and formulate this as a theory about themselves and whether such a self-description grasps their social function accurately is an entirely different question. Their differentiation does not depend on all this. Theories of reflection occur in such systems only secondarily – only for the defense of their autonomy and only on the basis of the demand for meaning that the system already presupposes structurally. This has been confirmed in a remarkable way by the development of scientific theories, political theory, economics and by legal theory since the second half of the eighteenth century.

12 Because function systems are not differentiated as regions of being, collections or by means of unified viewpoints but instead by means of differences, a high degree of reciprocal dependence is possible. Such dependencies are often interpreted as constraints on autonomy if not as symptoms of the reversal of differentiation. Actually the contrary is the case. Functional differentiation promotes interdependence and an integration of the entire system because every function system must assume that other functions have to be fulfilled elsewhere. This is the precise purpose of the binary code: to differentiate its own domains of contingency and its own procedures for creating differences through differences – and not essentially for differentiating exclusive orders of existence. Operations can therefore switch very quickly from the legal to the political

or from the scientific to the economic code. This possibility does not deny system differentiation. Instead it is attainable only on the basis of it.

My position is that binary codes having these characteristics occur in social evolution and that, if they are put into operation, corresponding systems tend to be differentiated. Traces of such a development can be shown for ancient Greek culture in the clear differentiation of the semantics of logic and epistemology, politics and ethics, and economics and philia.[12] At first the traditional models of social differentiation according to the distinction of city/countryside and stratification predominated and society presented itself within a scheme of religiously justified moral communication. In this sense, both urban life and, subsequently, the life of the nobility are ethical postulates.[13] Not until modernity did society gradually switch to the primacy of functional differentiation that, since the middle of the eighteenth century, has been accompanied by a corresponding problem-awareness. Ever since, a plurality of functionally specified codes has steered resonance to the environment instead of a socially unified or upper-class 'ethos', and these codes lack integration only to the extent that a positive valuation in one code, for example, 'true' does not automatically entail a positive valuation in the other codes, for example, as legally or economically significant.

9

Codes, Criteria, Programs

The thesis of the functionally specific coding of the operations of modern society is only a first step along the path toward a more concrete description. We must always keep in mind that not all the system's communications are ordered in this way, i.e., divided into one or the other code. Differentiation never occurs as the decomposition of a given set of operations, but as the separation of subsystems operating under the direction of a code within society.

Binary codes begin as different, highly abstract schemas that, at the same time, leave unclear how the operations of society are actually regulated. At first glance they appear as the coding of preferences. This would mean that truth is better than falsity or legality better than illegality or that it is better to have than not to have. But if the actual operations are observed along with the preferences that are expressed in them something different is revealed. The truth of the proposition that mice have tails is not valued as highly as the demonstration of the falsity of important physical theories. Within the legal system much effort has been expended on shifting certain laws into the domain of non-constitutionality (or in other words, there is no preference *of* particular laws or statutes *for* constitutionality), and the same is true for the economy. Many firms would be in a more fortunate position and would have better business results if they did not own certain plants.

In order to deal with such situations theoretically two levels have to be distinguished in the analysis of system structures: the level of coding and the level on which the conditions of the suitability of operations are fixed and, if necessary, varied. Or, to repeat an argument of the preceding chapter, the code's values are not criteria. Truth itself, for example, is not a criterion of truth.[1] Criteria *refer* to binary coding, according to the established tradition of concepts like *canon, criterion, regula.*[2] But they *are not* a term of the code itself.

We will formulate this difference of levels with the distinction between *coding and programming.*[3] On the level of coding a system is differentiated by means of a binary scheme. At the same time it establishes itself on this level as a closed system. This means that a value can be abandoned only for the sake of its counter-value. Accordingly, the alternative 'true or false' is permissible, but not 'true or ugly'. The codes are closed 'contrast sets'.[4] Programs, however, are given conditions for the suitability of the selection of operations. On one hand, they enable a 'concretizing' (or 'operationalization') of the requirements that a function system has to satisfy, and on the other, they have to remain variable to a certain extent because of this. On the program level a system can change structures without losing its code-determined identity. On this level, therefore, learning capacity can be organized to a certain extent, so through the *differentiation of coding and programming* a system acquires the possibility of *operating as closed and open simultaneously.* As a result, this differentiation, together with its accompanying capacity for articulation, is *the key to the problem of social resonance to the exposure to environmental dangers.*

Viewed historically, this kind of differentiation developed very gradually and became a necessity only when function systems were sufficiently differentiated. Within the existing tradition of political ethics and natural law the code-values (positive/negative) and the generalized formulas for the conditions of the suitability or usefulness of behavior could not be distinguished. Instead, the unity of the good and the right was rooted in a religious semantics that transcended the domain for which the difference of good and bad was meaningful: the world. 'The good' is thereby doubled. Transcendently it operates without a counter-value, but in the world it operates with the counter-value of 'the bad'.[5] It becomes logically ambiguous and forces the construction of a multi-level theory, for example, a hierarchy of laws. Nevertheless, Enlightenment thinkers like Rousseau followed this scheme even if 'the good' was sanctified as nature and made its ambiguity felt only when it had unfortunate consequences. The French Revolution demonstrated this in a striking manner and with it brought the history of wisdom to an end. All reflection then has to start *ab ovo* taking this fact into account.

A kind of demoralization of the most important codes – resulting from experiences with a monetary economy – had already been attempted.[6] Correspondingly, the history of economic reflection began with a search for a functional equivalent for morality, namely with Adam Smith's *Wealth of Nations*. By then science had long since promoted the idea of an 'invisible hand' to show that scientific progress was near and that it would bring immeasurable good with it.[7] Of course, there was no reason for the fear that things can go wrong which haunts us today. The metaphor of the 'invisible hand' and the reference to successful progress was therefore enough to refute the theories of transcendence which in those days were rejected as 'dogma'.

For the history of ideas we now find belief in progress with its supporting metaphysics in a period of transition. The new order of functional differentiation opens up possibilities that could not develop according to the levels of the old society. An entirely new kind of theory of reflection becomes possible – one related to the autonomy, self-value and function of the individual function-systems without considering their interplay. Whatever is triggered structurally is, at the same time, still screened off semantically. It has not yet made its presence felt and becomes acute for the first time in the nineteenth century as the 'social question'. But social reflection found no clear guidance from this and, lacking any consideration of environmental problems, no external support.

Even if one defers the question of ecological problems for the present, it can still be assumed that functional differentiation – if it develops as a consequence of the differentiation of codes – leads to new ways of formulating problems and new theories of reflection. In this regard two things must be kept in mind. The first is that the levels of coding and the programming of function systems become more clearly differentiated. The second – and this compensates for the differentiation – is that programs determining the criteria of suitability are formulated only in co-ordination with particular codes and are not transferred from one code to another.

What this means for theories of reflection that try to describe the system's unity can be seen with the help of the following diagrams of the legal system. The development of a regulative semantics goes from the hierarchically inclined composition in figure 1 to the more rigorously differentiated composition in figure 2. If a hierarchy remains then it can reside only in the subordination of the programs to the codes. Accordingly, the theory of law abandons the medieval interpretation (predominating in canon, Roman and even in English law) that decisive authority is given only for legal operation and has no authority in the case of illegal. In the nineteenth century, however, the

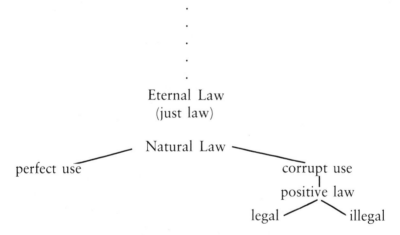

Figure 1 Hierarchical construct of the legal system

mere fact of authority was the goal. But if this happens the legal system is, as it were, coupled to the political system without any awareness of its own function. Thus its unity cannot be reflected adequately. The transcendent, external support supplied by natural law is eliminated and (for the time being) not replaced. So the use of natural law as the theory of law has to be rejected within the legal system. Programs assume the task of the correct handling of the values 'legal' or 'illegal'. The unity of all the

	code	program
unity	?	justice
operation	legal/illegal	valid legal norms

Figure 2 Differentiated construct of the legal system

conditions of the correctness of programs had been reflected traditionally in the title of 'justice' (by which a virtue or general norm was understood). With a more rigorous differentiation of codes and programs directing the operations the theory of law had to make its concept of justice more precise.[8] But it still finds itself in the embarrassing position of having to formulate an expression of unity for the difference of legal and illegal that, as a meta-norm, idea or ideal no longer fits the semantic domain of justice. The allowance of greater contingency in system differentiation and reflection on the structural and semantic level leads to the still largely unsolved problems of a satisfactory self-presentation of the system, to say nothing of a theory of the encompassing system of society.

One can assume that this provides a chance to extend the resonance capacity of society and its function systems. The metaphysico-moral conceptual framework may have been too narrow a context of possibilities. Although its rejection is not necessarily to be regretted any attempts at a 'rehabilitation' ought to proceed very carefully. All this requires further investigations. In any event we have to begin from the fact that resonance to the exposure to ecological danger is created essentially through these function systems and cannot be, or at most only secondarily, a matter of morality. To put it more precisely, the internal dynamics and sensitivity of function systems like politics, economy, science or law are disturbed by environmental problems. Sometimes this happens directly as when resources dry up or catastrophes threaten. But it also occurs indirectly via socially mediated interdependencies when, for example, the economy is forced to react to legal precepts even if it would attain better results following its own ideas.

On one hand, functional differentiation is possible only through the _rejection of redundancy_. Function systems cannot step in for, replace or even simply relieve one another.[9] All equivalences are ordered according to a specific function, i.e., within the system. Politics cannot take the place of science, nor can science the place of law – and so forth for all relations between systems. The old, multifunctional institutions and moralities are, therefore, dissolved and replaced by a co-ordination of specific codes to specific systems that distinguishes modern society from all those before it.

On the other hand, functional differentiation triggers an enormous *internal dynamics* within the function systems which combines intense resistance with very specific sensibilities to irritations and disturbances. But this impedes the theoretical description of the social system as a whole. Each function system has to be analysed individually according to its own specific resonance capacity. Nevertheless, social resonance as a whole is not merely the sum of the resonances of each of the specific function-systems. The subsystems are environments for one another. They can produce a process of resonating disturbances when one subsystem reacts to environmental changes and alters the social environment of the other subsystems. In this way a scarcity of resources can produce political problems in addition to economic problems like price increases. It can even force scientific research into certain directions instead of others. An extreme political sensitivity to environmental questions may burden the economy with additional costs, mean the loss of jobs and lead to political problems of its own, and this same political resonance capacity may trigger a new wave in the *flood of norms* overwhelming the specifically juristic mode of handling legal questions. This wave then surges back into the political system and it begins to operate schizophrenically, namely bringing more under legal jurisdiction (*Verrechtlichung*) when it wants to bring less (*Entrechtlichung*).

But disturbances are not the only things transmitted and thereby partially absorbed and reinforced. The working together of function systems is also necessary in practically all cases. For example, *scientific* research has made the construction of nuclear power plants *economically* possible through a *political* decision about *legal* liability limitations. The world is just not constituted so that events generally fit within the framework of one function alone. Functional specification is an effective as well as risky, evolutionarily improbable achievement of complex systems. It is purchased at the cost of extensive internal interdependencies that are controlled by system boundaries. But it would be quite mistaken to infer dedifferentiation from these reciprocal dependencies.[10] Instead they are a proof that modern society differentiates its primary subsystems with reference to specific functions and thereby prevents them from stepping in reciprocally for one another as functional equivalents. It also makes them depend on one another whenever problems can be solved only through their co-operation.

Society's resonance therefore has to be analysed on two levels at the same time (and here, as always, the idea of different 'levels' conceals a theoretical systems-problem). On one hand, resonance is conditioned by society's differentiation into function systems (instead of into levels with dismay in the lower and responsibility in the upper). On the other, it is structured by the different types of codes and programs of the subsystems that affect one another according to the general model of system and environment. As one can see readily, this produces effects within the system which are unlike the changes in the environment that originally triggered them. These in turn are observed with respect to their own degree of danger and so are in need of control. But all this occurs, if at all, only within function systems according to specific codes and programs.

10

Economy

Among society's many function systems the economy deserves first consideration. By the economy we mean all those operations transacted through the payment of money. Whenever money is involved, directly or indirectly, the economy is involved regardless of who makes the payment and whose needs are affected. This occurs, for example, in the collection of taxes or government expenditure for public goods. It does not occur in the case of pumping oil out of the ground except when this process is regulated economically to produce a profit that can be expressed monetarily.

This definition of the economy describes the modern one that is differentiated through the monetary mechanism.[1] Compared with earlier monetary systems like the medieval, however, it demonstrates a striking limitation on what can be bought with money. Today neither salvation nor the special providence of transcendent powers, neither political offices nor tax-rates, neither government assessments nor similar sources of income can be purchased with money.[2] This *restriction* is the indispensable condition of differentiating the economy from religion and politics. It is the condition of the *unification* of the economy and its *autonomous closure* as a self-governing, operative function-system of society. In other words, only through limitation does the economy achieve the immense internal complexity of a monetarily integrated system. In the same way it also increases its capacity for encompassing the satisfaction of needs and production and moves the traditional domain of the domestic economy further to the margin. Limitation, therefore, is the condition of its expansion, and this expansion contains the much deplored consequences for society's environment.

Originally the economy was coded through property. This forced *every* participant into the alternative of owning property or not regarding *all* the goods capable of being owned. Ownership by one participant excluded all the others. This was the only way that exchange was possible and property was capable of fulfilling certain ecological functions. A tolerant treatment of nature is beneficial to property because one can defend it against others and they can eventually be brought through legal channels to replace any damage.[3] In its pre-monetary form, property, especially as the ownership of land, was not capable of being differentiated sufficiently. It remained, for example, the quasi-necessary foundation of political power (feudalism). Only the binary coding of the economy through money and a change of the code having/not having to the code payment/non-payment led to the complete functional differentiation of the economic system.

Today, because of its monetary centralization, the economy is a rigorously closed, circular, self-referentially constituted system because it effects payments that presuppose the capacity for making payments (thus the acquisition of money) and manages this capacity. Thus money is a uniquely economic medium. It cannot be introduced as input from nor transmitted as output into the environment. Its exclusive task is to mediate system-internal operations. Since the system can negate, these operations consist of decisions. Payments qualify themselves against the background of the possibility of not paying (or not being able to pay). To pay or not to pay – that is, quite seriously, the question by which the existence of the economy is determined.

Code and programs are inevitably separated according to this basic operation. The code exists because it makes a difference whether someone has (the right to) certain sums of money or not. Only someone who has a certain amount of money and can part with it is in a position to pay because payment is the transformation of having into not having. The same is true, conversely, from the receiver's point of view. The code is the condition of the system's being set in motion and keeping it going and for the constitution of the system by events, in this case – payments. Such events (payments) are meaningless by themselves if reasons cannot be provided for motivating their performance for example, for satisfying needs or improving one's position for future payments. In this respect the system must be thought capable of learning, i.e., capable of reacting to changes in itself and the environment. Criteria of correct behavior, i.e., programs have to be created for this. Needs cannot be programmed directly. They occur and make their presence felt. But for the system they are environmental givens. The system remains dependent on a regulation of the operations internal to it, i.e., on a programming of payments themselves. This is accomplished by *prices*.

Prices permit a rapid determination of whether payments are right or not. This merely requires a quantitative comparison, and because this is so easy the question arises whether the prices themselves are right. So the programming requires a programming. Earlier economic orders (cf. figure 9.1 above) tried to do this with the theory of the just price. But in their case *just* meant 'corresponding to the market situation' while *unjust* meant 'interdiction of a profit that was not justified by this situation'.[4] The fixing of prices simply to improve one's own solvency or for commercial cunning was forbidden. Since code-values and programs were not distinguished this would amount to greed (pleonexy).

In the transition to modern society with its capitalistic economy this restriction disappeared and was replaced by restrictions internal to the economy. The legal system that interprets the justification of prices as contract reacted to this by removing restrictions on contractual freedom.[5] But this immediately affects only the operations of the legal system itself, for example, the ability to sue for payments. For economic calculation, prices are regulated by the economic process itself and require no external regulation (for example, natural law or morality). Prices are determined by what people will be willing to pay in the market and this is determined by the money supply available. The theory of the just price was therefore abandoned in the eighteenth century when it became evident that the economic system itself places restrictions on the profit motive and commercial cunning.

The system's coding and programming became a purely internal matter, while the environment placed restrictions on the system which were expressed within it as prices and price changes. Peace-making and the peace-promoting function of trade is an often repeated theme in the seventeenth century. But at the same time the securing of peace was left to the state and the international balance of power even if the economy could enforce the prevailing market prices without having to bother itself with the preservation of peace when it determines prices. Politics and the economy are functionally differentiated systems. Therefore, if politics intervenes in the process of price formation (which, as is known, happens to a great extent) it transforms economic problems into political ones. But the difference between the two systems remains.

A basic and very far-reaching feature of this system and its form of resonance is commonly designated with the concepts 'market' and 'competition'. An adequate market theory does not exist, even in economics. What is noticeable is a great amount of differentiation among contexts of competition, exchange and co-operation. Viewed historically and sociologically, this is a highly unusual structure. At present, co-operation and exchange normally do not occur among those who are in competition. This makes it possible to keep the competition of interaction and communication open among the competitors and to reduce it to a mere calculation of the social dimension of everyone's respective behavior. As long as the system determines itself through competition it can spare itself the difficulties and ramifications of the conjunction of interactions. But in doing so it has to renounce the possibilities and certainties of control that accompany direct communicative agreement (whether these turn out to be positive or negative). The system reacts via 'contagion sociale',[6] via a quasi-simultaneous processing of expectations *vis-à-vis* the processing of the expectations of others. The 'double contingency' embodied in this does not lead to the formation of systems. Instead, it is freed to make decisions, under the conditions of uncertainty, about the chances of success that are contingent on the decisions of others.

One of the most remarkable consequences of this – that more than anything else determines the resonance capacity of the economic system – is an extraordinary increase in tempo. The system operates so fast that it is limited to observing events and cannot integrate them through structures any longer.[7] Every intervention therefore acquires the character of an event, an impulse, a provocation, a stimulation or destimulation of changes in the system itself, and the unforeseeable effects of this continually produce new impulses of the same type.

How, under these circumstances, a directive intervention is possible within the organized complexity of the market is something that would require a detailed investigation. Above all, the problem exists in the money market, which varies in almost complete disregard of the environment but influences all the other markets. Yet the quantitative dimensions of this market – fluctuations of up to several hundred billion dollars daily – give a person pause.[8] Whoever, in view of such phenomena, still wants to appeal to an environmental ethics has first to be concerned with the financial instruments of this ethics.

One has to admit that this (turbulence-emphasizing) description of the system is by no means complete. This calls for a closer look at how payments and the operative autopoiesis of the system are actually disposed of. The economy can be called 'capitalistic' only to the extent that it connects payments with the reproduction of the capacity to make further payments, above all from the point of view of the profitability of investments. Capital is necessary because there is a time-lapse that has to be bridged between any payment and the reproduction of the capacity to make further payments. Revenues are not yet available when the means of production are purchased (for example, in the case of seed-corn) or are left unsold. The more available capital is, the greater and more indirect are the relations of production and consumption that can be included in the economy. But in the final analysis this means that capital investment must be calculated economically, i.e., rationalized in terms of the preservation, reacquisition and increase of capital.

In a money economy (an inevitability in modern society) a capitalistic self-control exists only as a possibility. If this is ignored and, for political reasons for instance, unprofitable investments are made then someone has to assume responsibility for this. Payments are made without reproducing one's own capacity for making further payments. In this, so to say, counter to the money-cycle way, the incapacity for making further payments is passed on to others and they are forced to make unprofitable payments (perhaps in the form of taxes). They have to regain their capacity to make further payments then in other ways – perhaps through increased prices. In the same way, private households are forced to pay for goods that are consumed immediately and do not reproduce the capacity to make further payments. So even the private household is removed from the capitalistic sector of the economy and made incapable of reproducing its capacity to make further payments if it does not look for other modes of income, for instance through labor.

This 'double cycle' of the capacity and incapacity for making payments results directly from the idea that an economic system consists of payments. Payments themselves are twofold: they produce in the payee the capacity for making further payments and in the payer the incapacity for making further payments (with this money, of course). But such individual events are possible only in a dynamic system, i.e., only under the condition that the capacity and incapacity for making further payments can be transmitted or passed along. The metaphor of the 'cycle' means nothing more than that, in any individual case, it has to proceed in the hope of 'continuation' and that no system operation can escape its inexorable law. The 'identity' of the capacity to make payments is conceivable only on the levels of systems – but not in such a way that the very thing that one gives up is returned (circulated) back to the payer. The metaphor of the cycle represents the unity of the system, and this means the autonomy of the system's autopoiesis. The reality resides in the conditioned operations themselves.

Under such conditions the economic system has to look in two different directions at the same time to provide for the preservation and reproduction of the capacity to make payments: to the left and to the right, as it were. On one hand, the issue is profitability and on the other it is the supplying of the economic conditions for the fulfillment of public duties and the provision of work. The credit mechanism provides for a certain margin by creating the capacity to make payments precisely where this would not occur as a result of circulation alone.[9] This can be steered to a certain extent by central banks that are always in the position of being able to make payments. But it is questionable whether there are criteria for this in the system – apart from the intention of normalizing the system's present perspectives of the future.[10] In any event, the mere necessity of the safeguard of a central money supply (which is nothing more than the arbitrariness of the transition from the incapacity to the capacity of making payments by the central bank) is still not a criterion for its use. Even theoretical or political guidance from equilibrium theory or multivariable models of optimization is simply the elimination of the tautology of economic self-reference: an interpretation of the history of the system (i.e., data) so it can continue to be written.

More important than the structural metaphor of circularity is the fact that, since payments and the regeneration of their possibility are events, *time has to be built into the system*. This means that the economy is continually concerned with gaining time and forming capital so it can have time available at all times. Thus the system develops its own future/past perspectives, its own temporal horizons and temporal urgency. One cannot simply assume that this system-time accords with the temporality of the processes in the ecological or even the social environment of the system. For this reason there is limited resonance capacity. Even if, for example, fossil fuels deplete rapidly it may 'still not yet' be profitable to switch to other forms of energy. This time-loss is what concerned ecologists regret most of all. Even if the gradual exhaustion of resources or a forthcoming election can be significant to economic calculation – decisions in the economy are made according to its own conditions.

One might think that preserving the time-consuming double cycle of the transmission of the capacity and incapacity for making payments would be enough to keep the system busy. Therefore, resonance for environmental questions is possible only when the exposure to ecological dangers is brought into this double cycle, whether this comes about because one sees a possibility for making a profit in them, opens new markets, produces new or transferred incentives to buy and especially because one increases prices and forces them on the market. It can also come about because one makes unproductive payments, increases the incapacity for making further payments and passes this on. The economy has to realize both possibilities, and it can reinforce both of them when it puts more money into circulation. Whether it can still fulfill the (patently fictive) expectations of its own theory, for example, attain an equilibrium or optimize a welfare function – is more than doubtful. It is truly astonishing that, to a large extent, the constant (1) selecting of profitable payments; (2) financing of public expenditures; and (3) providing of labor succeed over a very wide range of goods and needs.

Accordingly, the concept that the economic system produces for itself of the ecological environment (as distinguished from humans and society) is limited by the possibilities of adding its own operations to it. In this sense, Dieter Bender defines environment, 'as the totality of all naturally provided, non-produced goods and services that provide streams of profit to the individual participants in the process of production and consumption'.[11] Although this definition suggests a direction of flow it includes the absorption of economic declines because the economic system draws profits from this absorption too. But right from the very beginning this definition is calculated towards compatibility with internal economic operations and not to the particularity of the environment. Moreover, through precipitous equalization it conceals the largely typical problem for an environmentally minded economy of separating the problems of levels and amounts from those of allocation and deciding each separately.[12] None of this is a shortcoming or a constriction that should be criticized. Instead, it is the condition for the system's ability to steer itself internally according to its difference from the environment. In the same way it also provides the limits of possible resonance.

Only to the extent that the environment is brought into the economy in this way and internalized with the help of quantitative or profit calculi can there be economic motives for handling the environment with care as the property theory of the physiocrats had intended. Resonance to environmental data and events is then regulated through prices and what influences them. On one hand, prices are a critical instrument in the discovery of environmental opportunities. When prices rise so do the opportunities for the increase of production, including the extraction of material and energy from the environment. When prices fall activities that are no longer profitable are discontinued. Small profits stimulate production too, even if they are accompanied by distant, unperceived (by the market) risks of catastrophic consequences, and even if businesses felt a sense of responsibility for such consequences it would still be rational economically to leave these out of consideration. This comes under the much discussed asymmetry of internal advantages and external liabilities. But it also demonstrates that it is not always possible to solve the problem through forcing the internalization of costs.

On the basis of theoretical models the economic theory representing the position of the system's self-regulation provides relative optimism for the possibilities of ecological adaption.[13] It assumes that self-regulation is determined by prices alone and that this enables the best possible distribution of information about the environment. One could also say it is determined by the prices that result from demand or, somewhat more aggressively, by the prices that the market will allow. But this is, as mentioned previously, a system-internal theory of system-internal processes that, so far at least, believed production was expandable according to a price-expressed need. Environmental conditions are considered only as constraints on what is technically possible and economically profitable at any time.

Within the conceptual framework of economic theory it is possible to determine that the marginal utility and marginal costs of the protection of the environment ought to balance out[14] and thereby derive a principle that both makes the resonance capacity of the economic system possible as well as limits it. But enormous problems of measurement and practical problems of attribution remain. Above all, one must realize that the decisions of the economic system never decide for the whole system. Instead, they are guided by the 'internal' environment of the economic system, namely, by the market.[15] But the latter is prefiltered to such an extent that an all-encompassing economic decision rule directed at the environment would find no application. It is also difficult to imagine that prices could be so manipulated by an external, politico-legal dissemination of data that the subsystems would decide about production and consumption as if they were guided by ecologico-economic marginal utilities. If such a need for regulation is taken as the point of departure – and this might very well be accepted today – then it suggests that the political system fixes amounts (above all, the amount or level of acceptable environmental pollution or the amount of the final consumption of irreplaceable resources or even negative amounts in the form of costs) and that the economic system concerns itself with an optimal distribution and use of these amounts.[16] This seems compatible with a market economy. But is it?

In order to take this under consideration and eventually to solve it let us go back to the general code-paradox of the economy, the paradox of scarcity. It states that the elimination of scarcity through the appropriation of scarce goods increases scarcity. This paradox is disguised (the invisible hand) for the market through widespread economic success and especially through growth. The so-called 'external costs' are used to accomplish this. But when one establishes a difference between amount decisions and allocation decisions this establishes, *instead*, a different, functionally equivalent form of paradox elimination, namely, a self-difference, a hierarchy. The market had served to conceal the difference between these kinds of decisions. Through the establishment of a hierarchy the difference becomes evident. This also brings with it the typical problems of hierarchical paradox elimination, and what Douglas Hofstadter calls 'tangled hierarchies' or 'strange loops'[17] are encountered in many places. One intends to operate on one level and unexpectedly the operations are on another. Decisions about amounts interfere with the allocation process and create compelling grounds for changing the pretext of the amount due to the kind of allocation.[18] But this does not mean that this is bad or that it leads to insoluble problems. No system is destroyed by logic. One has to realize, however, that a different market-strategy is being used to eliminate the paradox; one that, to a great extent, exposes structural contradictions and decisions.[19]

Finally, we have to consider entirely different kinds of theoretical ideas, those guided by input/output models. We have to discover externalized costs and reincorporate them in the economic analysis. We also have to reveal the environmental consequences of economic activity and put them into a form that allows decisions to be made. So one part of the economic literature requires that the goals of the economic system should be extended to the ecological side-effects of economic activity.[20]

In connection with this a distinction has to be made: the economic system itself has no goals because, as a closed autopoietic system, it is not directed toward any output. The best that can be said is that production organizations observe environmental protection as a secondary goal, especially when they are directed by managers and when they do not perceive an urgent reason for a market-price guided dividend policy beneficial to shareholders. Similarly, a consumer – if ecologically-minded – might be willing to pay more for environmentally safe soap. Behavior can be modified in an ecologically desirable sense in the economy – but not without affecting production costs, and consequently taxes and preferences for goods.[21] It may make very good sense to speculate on ecological enlightenment, improved clarity of causal connections and on 'changes of mind' or 'value shifts'. But it is impossible to tell, at present, how such changes will work out in the economy and what still-unknown side-effects they will trigger.

This self-referential type of economic information processing leads to transforming problems into costs.[22] They then become parts of the calculation that decides whether it is economically rational to make the corresponding payments or not. In this form the system distinguishes between solvable and unsolvable (not financable or financially unprofitable) problems. Both the results and the prospects themselves remain problematic on the whole.[23] The results remain problematic because the rejection of the payment of costs transfers the unsecured amount into the cycle of the transmission of the incapacity to make further payments and can overburden it. The prospects remain problematical because the definition of the problem does not reveal all the aspects of the problematic through the concept of costs and the schema of paying/not paying. Just as with conditional legal programs this is a matter of a specific technique for dealing with higher structured complexity – of a very efficient, barely improvable technique based on a one-sided selection of its starting-point. Here as well as elsewhere the solutions to problems are not without consequences. Therefore whether the economy solves its problems or not problems remain for other domains of society.

Because of the highly selective resonance of its object domain a theory of the interconnection of price, costs and production is not capable of judging our society's exposure to ecological dangers. It is not even capable of giving political advice on this question. But it provides a good idea of the self-determined resonance capacity of the economic system and its self-referential closure. Formulated as a principle it says that whatever does not work *economically* does not *work* economically. This is correct *mutatis mutandis* for all the other systems too, even for politics. For the economy, the question will always be with which prices will the capacity to make further payments be passed on and how can the incapacity to make further payments be transferred. This is the only mechanism that combines autopoiesis, resonance to the environment, continuation of production and the inclusion of an unintelligible, noisy environment in this process.

The key to the ecological problem, as far as the economy is concerned, resides in the language of prices. This language filters in advance everything that occurs in the economy when prices change or do not change. The economy cannot react to disturbances that are not expressed in this language – in any event, not with the intact structure of a differentiated function-system of society. The alternative is the destruction of the money economy with unforeseen consequences for modern society.

This structural restriction to prices is, however, not only a disadvantage, not only a rejection of other possibilities; it *guarantees* that the problem, *if* it can be expressed in prices, *must be also processed in the system*. As always, the reduction and increase of complexity work hand in hand, and it is difficult to see how, without such a restriction of resonance capacity, the extensively compartmentalized, alternativeless possibilities of reaction to environmental stimuli could be produced. Otherwise the situation turns out to be just like that of a woodcutter who sooner or later has to find out that there are no more trees to be cut down.

11

Law

In the contemporary ecopolitical discussion the contrast between the language of prices and the language of norms is as striking as it is disarmingly simple.[1] This corresponds to the long-standing distinction of society and state and suggests a simple alternative in the reaction to ecological dangers. This leads immediately to the following consequence: whatever does not fit within the language of prices has to be expressed in the language of norms. Whatever the economy does not bring about on its own has to be accomplished by politics with the help of its legal instrument. Finally, the ecological problematic runs up against a residual political responsibility that unexpectedly becomes the all-consuming responsibility of a constant state of vigilance.

This alternative is formulated much too simply and leads in the final analysis to the fact that more is required from politics than it can perform. It leads not only to avoidable disappointments but also to an overburdening of the political system with unfulfillable demands and it is misleading. Politics is then plunged into verbal debates. It is forced to offer hasty, false solutions, defer problems or try to gain time, which inevitably leads to radical disappointments with it instead of the realization that this system is capable of resonance only within the context of the frequencies of its own autopoiesis.

We had already rejected the conceptual conditions of this account because it tends toward over politization and consequently to political fiasco. We replaced it with the argument of a functionally differentiated society. Above all, this means that politics as well as the economy are only subsystems of society and are not society itself. It also means that the function systems of politics and law have to be distinguished more clearly than usual. Admittedly, law stands in close relation to politics since law-making typically requires previous political agreement. It is a system that is sensible to political situations and, as a result, singularly capable of resonance. But at the same time it is a closed system that can create law only on the basis of law, norms only on the basis of norms and through its court apparatus makes sure that this condition of its autopoiesis is observed.[2]

The legal system receives its autopoiesis through coding the difference between what is legal and what is illegal. No other system operates according to this code. This binary coding of the legal system creates the assurance that if a person is in the right then the force of the law is behind him or her. Uncertainty about the law exists only in a form that can, in principle, be rectified, namely, in reference to decisions that can be made in the legal system itself. This assurance is attainable within society only if the legal system alone assesses what is in accordance with the law and what is not. Besides, only one system of society may use this code,[3] and since no other system uses it, the legal system cannot import or export what is legal or what is not. It tells only itself and not the rest of society (its social environment) what is legal or not in any case. The incontestable social effects of law rest on the occurrence of this *within the legal system.*[4]

Since in any case only one of these two possibilities exists for the legal system – there is no third possibility – the schema contains a complete description of the world.[5] For the performance of operations and the reproduction of standards all that remains to be determined is whether any particular case is in accordance with the law or not. In other words, standard legal programs must be provided that fix the conditions of legally correct decision-making. These can be found in laws or ordinances, statutes or procedural rules, in judicial rulings or contractual agreements. On the level of programming the system is closed and open at the same time. It is closed because a norm can be obtained only from norms themselves (no matter how logical the mode of inference or arguments may be judged). It is open because cognitive viewpoints also play a role in this. Cognition is required both for the system's environmental orientation as well as for its own orientation, both for the determination of the empirical conditions of the application of norms as well as for judging the adequacy of or the need to change the norms themselves. The system operates completely 'open' for environmental conditions and their possible change. In other words, it can learn. But first the autopoiesis of the system has to be used, i.e., it has to proceed according to the difference of what is lawful and what is not according to legal programs – even if only because otherwise it would never be possible to know that a legal process is at issue.

Thus the code is also the autocatalytic factor driving the system, both in terms of its need for supplementation and in the formation of highly complex program structures. As internal structures these programs work only on the internal processes of the system. The system can know its environment only through them – even if this is as a disturbing noise that can be corrected only by a modification of the programs. This corresponds to the social function, i.e., the socially internal function, of law to take precaution against conflict and also to provide for stable expectations in the case of disappointment. In this case the social system's environment comes into consideration at best as the occasion for conflicts (and consequently as the occasion of precaution against conflict). The environment disturbs the smooth, customary fulfillment of expectations through 'noise', and in many ways property functions to transform this eventuality into a partially economic and partially legal problem.

Therefore within the legal system the basic figures of legal thought represent a *social* desire for order (and in such a self-evident way that it almost goes unnoticed that this is a structural selection, i.e., a selection out of a plurality of possible ideas of order). This holds for all figures of reciprocity, exchange and the distribution and generalization of the relevant valid conditions, including the categorical imperative. It also holds for the semantics of freedom and the restriction of freedom ever since the bourgeois revolution. Legal ideas of order apply to socially *internal* relations, and it would surprise jurists to think that someone thought it necessary to advise them about this.

But it is precisely the ecological debate that provides the occasion for this. For even law reacts only in its own system-specific way to the exposure to environmental dangers, and there is no guarantee in advance of an adequate proportion or a causally successful reaction to the danger. The form in which law programs its code, i.e., translates it into the conditions of correct action, is fixed in the future-perfect tense. It imagines an action as completed in the future. This is the basis of the profound relation of law to freedom. Events occur on their own. The law simply states how they are treated if they occur, therefore the basic form of law remains the conditional program, no matter how much talk there is in environmental law about 'goals'. The law can never attempt to capture all causal factors successfully or to determine all processes legally. Of course, one can decide for reasons of a political goal to make laws and to justify law-making as a means to the goal. But it would be a mistake to see a causal statement in this. Whatever degree of probability the law contributes to the attainment of the political goal is a question that depends on *other* factors. Legally, this is a matter of the clarity and univocality of the answer to the question of what would happen in the case of a conflict and of the possibility of forming expectations for this.

This form of legal regulation can be proclaimed as the protection of freedom, indeed as the promise of freedom. Viewed more prudently it is a matter of a specific technique for dealing with highly structured complexity. In practice this technique requires an endless, circular re-editing of the law: the assumption is that something will happen, but how it will happen and what its consequences will be has to be awaited. When these consequences begin to reveal themselves they can be perceived as problems and provide an occasion for new regulations in law itself as well as in politics. Unforeseeable consequences will also occur and it will be impossible to determine if and to what extent they apply to that regulation. Again, this means an occasion for new regulation, waiting, new consequences, new problems, new regulation and so on. Presumed foresight is, therefore, an important auxiliary motive that keeps the process going. It results in an extreme complexity that is comprehensible in the legal system only historically.

Even in handling ecological problems law is bound to its own function, its corresponding technique for reducing complexity and, especially, to its own distinctions. All of these co-operate in programming the reduction of complexity. One can see this even more clearly as the machinery of environmental law is already in full swing.[6] Thereby, problems result both from the historical presence of law (without which law would not exist, could not even react) as well as from its sociofunctional specification.

When the ecological problem was first raised not a single area existed for law as unexplored territory to be formed and covered with a net of new regulations. There is only one legal system of society and this is always present. Law itself is always completely formulated and can only be changed. Consequently, 'environmental law', accompanied by new kinds of formulations of the problem (or with easily revisable formulations), cuts into customary legal domains like area planning law, legal competence, police law, business law, tax law and constitutional law.[7] As long as such a co-ordination does not succeed innovations remain abstract. They remain merely problematic ideas. The presently topical 'testing of environmental compatibility', for instance, looks like a new concept, like a legal innovation. But it is nothing more than the idea of a better harmonization and possible adaptation of existing regulations to different legal domains, understood and reorganized to cut across existing systematizations. All further development of law is bound to corresponding points of departure and has to take into consideration that other regulations remain in effect. Otherwise the idea of a relatively consistent, non-contradictory treatment of the legal code would have to be abandoned.

It is equally apparent that law can be developed only as a social regulative. The valid guiding principles for this may change, for example, from reciprocity through contract to the weighing of interests and the protection of trust or through numerous new legitimizations for the rights to and limits on freedom (whereby the need for limits can lead to the creation of freedom just as freedoms can reveal the necessity for their limits). This is why the distinction of freedom-rights and legal coercion have become the dominant formula for legal discussion about the environment,[8] although the distinction itself does not refer to the environment as such.[9] We are far from acknowledging rights 'to the

environment' *vis-à-vis* society, far from granting rights to trees
or for punishing dioxin for its toxicity by burning it. Instead, we
face the problem that because of a widely compartmentalized
system of subjective rights, we not only individualize the
disposition over a domain of our own (our 'own house') but at
the same time, under changed conditions, we leave the estimation
of one's own interests and the willingness to assume risks,
including the willingness to sell this willingness, to the bearer of
subjective rights. Of course, the law can make counter-precautions
through limits on freedom. But the question is, what does this
granting and limiting – under the control of the constitution and
normal laws – produce within society when it tries to create
resonance to the exposure to ecological dangers?

If the reaction to environmental problems must be met with
an internally goal-directed conceptual apparatus then an essential
incongruence of legal categorization is to be expected. Despite
the proven learning capacity of the legal system, its laws and its
dogmatic theories, this incongruence cannot be rectified, for the
regulation of communication in the system is something different
from its reactions to changes in the environment. Like every
system, the legal one is capable of resonance only in accordance
with its own structures.

A sociologically more important indicator of this is that the
*component of arbitrariness in environmentally related legal
decisions increases significantly.* This holds in at least three
respects:

1   for the necessity of defining marginal values, thresholds and
    units of measure for which the environment supplies no
    determination;

2   for determining the system's willingness to assume or tolerate
    risks when their transgression brings with it protective
    measures and eventually those that discount the costs or even
    proscriptions;

3   for fixing preferences regarding the extremely diverse conse-
    quences of environmental changes that are in large measure
    blocked, scattered and concealed by the price mechanism or
    even for the protection of concerned interests that cannot be

co-ordinated immediately because of the indirectness and opacity of causal relations.

All of these are by no means new types of problems for the legal system. But they acquire a new intensity and scope when a new ecological consciousness of the problem begins to affect the law. Natural law ceases to function precisely where it concerns nature, and even consensus (a kind of ersatz natural law) seems unattainable. At the same time it becomes questionable whether problems of this kind can be analysed, factored and ultimately solved satisfactorily with the standard legal method of handling cases.

Along with an increase in arbitrariness therefore, we find apparently contradictory, empty formula-like obscurities. These leave all decision problems open and merely give the impression that something is happening, at least on the verbal level. We will cite two examples of this without holding it against their author since they are typical of many others. 'Insofar as there is some room given to individual policy-makers for development it is recommended to concede to environmental protection a certain primacy'. Also, 'Their [the administration's, N.L.] task is to bring about a balance between the general public interests and individual concerns and to reconcile these with the requirements of environmental protection'.[10] Formulas of 'equalizing', 'balancing' or 'proportionality' can be achieved only arbitrarily. If the law has to resort to such formulas then a technically informed arbitrariness is not the worst solution. It is just not a specifically legal one.[11]

The transition from empty formulas to arbitrary distinctions is what makes it probable that any argumentative beginning in the usual sense of legal practice reintroduces these problems in more manifold and magnified ways. Such possibilities make their presence felt most clearly in dealing with risks. The standard rule of maximizing anticipated profit with a minimum of risk fails.[12] Anyway, it works only in the few cases where no uncertainty exists regarding the probabilities. As a general principle it is too risky.[13] Empirical research has shown that the willingness to take risks very often includes individual personalities, social systems, circumstances and previous experiences.[14] Therefore, maintaining any threshold of tolerance can be achieved only arbitrarily. Very often risk is valued highly and sought.[15] Moreover, the subjective factor in the estimation of improbability increases with an increase in improbability.[16] This is a problem that involves complexity too: distinct willingnesses to take risks cannot be added together and, to a great extent, depend on voluntariness so that they would be changed, if not nullified, if they were required by law. Regulation can occur but only arbitrarily and not without changing the consensus situation by the regulation itself. This means that centrally ascertained and established risk estimations and tolerances are unavoidable.[17] They cannot be based on consensus but have to be exacted, which then automatically reduces the willingness to assume risks.

Of course, the jurist never raises the question of how people come to estimate risks. Empirical research about risks is just as irrelevant as models of rational decision-making. One has to make decisions in accordance with maxims that one has ascertained personally. The fact that any residual risk ought not to be avoided, i.e., has to be accepted, is therefore acknowledged in practice as well as in theory. The justifications for this, of course, are not of much help. 'Social adequacy' cannot be inferred from 'unavoidability'. But even the appeal to an ethics that, for all practical purposes, is equally helpless, is no longer of any assistance since this too has to turn to reason and reason must then 'help itself'.[18]

Another consideration that remains unresolved concerns the 'balancing of goods' as the expression of ethical responsibility.[19] With this, the function of law to safeguard expectations in cases of conflict is not yet fulfilled. The rule applies to decisions in individual cases, i.e., it merely states that the courts will make decisions only after a careful examination of the situation. So, in the final analysis, the process of argumentation merely distributes the self-validation of legal decisions over many stages.

One could attempt to determine the probability of the occurrence of desired or undesired consequences. In this way legal regulation can extend to the choice of test procedures and the statistical control of the probability of the positive or negative determination of errors.[20] If a satisfactory determination of probability can be provided then uncertainty would no longer exist in the language of decision theory, only risk in the narrow sense.[21] But this would not solve the problem of the acceptance of risks, which would only be formulated more clearly. The advantage of consensus that exists when everyone begins from a different estimation of probability or when the entire question is left indeterminate would no longer apply. This would result in an unattainable agreement concerning whether positive or negative risks with a determinate probability are acceptable or not.

It does not even help if this question is made to depend on 'trade-offs' or on the level of expected profit or loss. First of all, no unanimity will be attainable in their estimation and secondly, profits and losses are distributed unequally in society.

One may also consider testing the readiness to assume risks in the market. At present there exists a clear inclination to proceed in this direction.[22] This happens when there are sufficiently localizable connections between profits and possible losses – for example, when in the case of childrens' pajamas with a high resistance to fire one cannot exclude the possibility that they also may be cancer-causing. Then a duty to reveal the risk may be sufficient. But for diffuse risks that are supplied indirectly through the environment, other forms have to be found that are directed toward the economic decision of the producer. Taken to its logical conclusion, this would mean that not only those who create risks for others should be made to pay for them or be obliged to provide insurance against them but also that those who live in danger must receive compensation for this – that the price of land in the flight paths of airports or in the vicinity of nuclear plants does not decrease but is kept constant or increases in compensation for living in constant anxiety.[23]

It is already evident that economic limits on the provision against risks of fatality are generally accepted *for others*. But it is hard to accept that one's own risks of fatality are accepted against payment. If ethical and eventually legal decisions diverge here then this is simply because in the latter case a decision (i.e., freedom), is demanded of the candidate of fatality. Moreover, such freedom would be economically exploitable *vis-à-vis* passive exposure. One could negotiate and increase the price. To the extent that the willingness to accept risks is indemnified the aversion to risks would be profitable – at least as a position of negotiation. In addition, this solution presupposes an antecedent identifiability (for oneself and for others) of winners and losers that is not guaranteed in most ecological risks because they are distributed too widely and indeterminately. Finally, the legal system is faced with the question of whether it is justifiable to hold someone who has agreed to accept a risk to a preceding agreement in the case of catastrophe – '*Tu l'a voulu, Georges Dandin*'. If not, then should not the possibility of being released from the agreement be included in the calculation from the beginning?

These problems – accompanying the very first steps in problem management – are added to those connected with the analysis, operationalization and factorization of problems. Included in these are disputes about the adequacy of measurement procedures, the power of experiments to provide evidence, and indispensable variables in simulation models. The problems of risk estimation will have to reckon with information that is constantly new in addition to changes in preferences. All that is needed is an atomic accident of the most trivial kind – and everything will have to be decided anew.[24] Further research may, and most likely will, demonstrate with a high degree of probability that when we apply rigorous standards we know less than we thought.

Additionally, we have to mention a problem of great practical importance that only touches our theme marginally. In the introduction of new or the exclusion of established technologies there are always direct social risks, i.e., direct social consequences and effects that cannot be assessed precisely and are triggered not by causes in the environment but from within society itself.[25] Here too – in the case of marginal risks – we find subjectivisms in assessments and the probability of improbable reactions as it were.

What does a concern with a 'rational' solution mean in view of such problems, especially where 'rational' means 'capable of or requiring consensus'? The structure of reason was directed at socially *internal* problems; developed towards problems of *social agreement*. One cannot ignore this for ecological problems either, especially when this pertains to social reactions. But the problematic *does not reside in the agreement*. It resides in the still-uncontrollable relations of system/environment. Accordingly, the classical model of finding a political consensus fails. Neither a liberal theory that would like to view solutions as the undisputed function of private decisions, nor a collectivist theory that thinks it will always know what the people want, offer convincing answers.

The typical jurist is satisfied by the idea that such questions 'have to be decided politically'. The recommended 'practical regulations' are only a variant of this idea. The third value that is excluded from the code of legal and illegal, i.e., what for the time being is neither legal nor illegal, appears in the legal system as politics. Thus the legal legitimization of political decision-making leads to the reintroduction of the excluded third value into the system. In this way the legal system makes use of a constitution and democratic legitimization to deceive itself that politics can handle problems better than law and that all arbitrariness can be transferred into this system for appropriate treatment and reintroduced as a legal norm. This stunning logical achievement of including the excluded third possibility may be admirable,[26] but it leads the political system, on one hand, to view law as its own instrument of implementation,[27] and on the other to decisions within the legal system that are not decided in a specifically legal way but are determined arbitrarily.

A closer analysis of these particular components of arbitrariness, even if they appear only within environmental law, show very quickly that there is no recourse to a 'nature of the matter' nor to a basic consensus of all those who think rationally and legally. This holds for all three decision problems mentioned above. The threshold values to be determined do not find a secure basis in nature. Ecological problems are simply too complex, interdependent, circumstantial, unpredictable, determined by the 'dissipative structures' of thermodynamic systems, the abrupt disturbances of stability (catastrophes) and similar structural changes, for this. The acceptances of risks – research in decision theory has come to this conclusion – are subjectively so different that consensus cannot even be reached in the case of assessments that agree. In addition, the scope and opacity of the causal connections imparted by the environment makes every value consensus trivial. The earlier rule of generalized reciprocity, the 'scratch my back and I'll scratch yours', and the categorical imperative fail and appear in the historical context simply as socially internal maxims. The same holds for the late rational-law attempts at developing rules for establishing a hypothetical consensus, procedural consensus and norms where consensus would be granted when someone behaved freely in a way that implied the recognition of these norms.[28] All this assumes that the problem has its roots in society and therefore can be solved in the social dimension. But the inclusion of the environment of the society in the genesis of social problems changes these problems essentially. Only a preventative consensus is attainable: an abstract agreement about preventing all possible damage in so far as the costs of prevention do not have to be accepted. In the meantime, cars race along the streets, lungs fill with tobacco-smoke, people borrow money, get married and risk criminal prosecution for tax fraud or antisocial behavior[29] – as if to prove that life without risks and without the assertion of highly individual preferences is of little value.

Under these circumstances the precipitation of political activity, social solidarity and the legal solution of environmental problems remains as abstract as it is inconsequential. A legal categorization of precepts for environmental concern can only be worked into the law, if at all, with other concepts. Dogmatic legal learning processes are slow and need decades, if not centuries, to boil down case experiences into concepts and maxims and to reformulate the law in terms of these. In addition, they assume, for whatever reason, that courts can be used to do this. In the case of environmental law this seems to be so only to a relatively small degree, if we measure it by the collection of regulations and the amount of literature.[30] The law is created for bureaucratic handling and seems to be designed for this from its mixture of arbitrary determinations and vague, empty formulas. Similarly, the administration of justice develops in connection with very heterogeneous areas of specialization. It remains to be seen whether this amounts to encompassing categorizations or even to the development of specifically legal modes of argument and justification. For the time being it is noteworthy that the legal system reacts to the desideratum of an environmental law with a considerable increase and complication of the regulation apparatus. Through the co-operation of the political and the legal systems a resurgence of norms has appeared at their boundary.

The political system finds itself in the need of having to profess and to cope with the desire to decrease [*Entrechtlichung*] and increase [*Verrechtlichung*] the scope of laws at the same time.

Finally, a last problem concerns the enforcement of law, its execution and the effective prevention of exceptions. Here research is concerned, on one hand, with answering the universal complaint about problems of enforcement in obviously insignificant and self-evident difficulties.[31] More important, however – because of the increasing attention given to environmental questions – is the appearance of new modes of enforcement that are difficult to fit into the law. These can lead to the reassessment of views and experiences and, at present, are identifiable only through tendencies of distorting existing institutions. On one hand, this holds true for 'private' efforts at seeing to it that public law is enforced which have to operate without the protection of subjective rights and are relegated to supplying information that forces public authorities to intervene.[32] On the other, enforceable law at present serves the administration largely as a negotiating position from which it can, in part, attain non-enforceable concessions, relinquish severity in enforcement and, in turn, put goals back into a gray zone of legality.[33] Obviously both exceptions to traditional legal structures can collide with each other when private legal pursuants cancel out or highlight legally problematic aspects of agreements. It may be that both of these new types of enforcement techniques can be brought into an equilibrium of 'checks and balances' and that courts can develop criteria for this. For the time being only a tendency to concede to administrations a greater scope in making judgements is noticeable.[34] One can observe, therefore, that ecological communication deforms classical structures of the legal system, and how it does this, on more than just the level of the content of norms.

12

Science

Perhaps we have been barking up the wrong tree. Perhaps it is not the economy or law but science or politics that is responsible for environmental questions in the differentiated system of society. Since politics repeatedly pins its hopes on science and establishes programs for promoting research and technological developments we will turn our attention now to the scientific system to examine its resonance capacity.

We are not concerned here with questioning the productivity of the natural sciences, or with examining and making a wholesale judgement about them. As before, we are interested solely in the systems-theoretical question of what is decided about the resonance capacity of function systems when they are differentiated through a specific code and when their coding and programming are distinguished according to this.

The difference between true and false is what matters for science's code.[1] Research programs are usually called theories. Even for these the differentiation of coding and programming, of truth-values and theories has the expected consequences: the concept of hierarchy as the form in which unity and rational connection are presented is abandoned and a differentiation of the scientific system into disciplines and subdisciplines takes its place.[2] After this, any possibility of determining the position of knowledge in reference to the unity of the system is relinquished. The disciplines work within a loose, expandable, theoretically (i.e., research-programmatically) non-integratable association that can be developed further, if necessary, through splitting-off, through subdivisions and new formations. The unsatisfied requirements of reflecting the unity of knowledge and its conditions of possibility resulting from this reflection converge in a new kind of theory of reflection of the scientific system, into an epistemology. The more pronounced the need becomes to distinguish between everyday and scientific knowledge the more the unity of the scientific system is reflected in this system boundary, the more exact the theory of reflection of the scientific system becomes. If a message about the meaning of the unity of the difference between true and false is ever to be found (see the position of the question mark in figure 2, chapter 9) then this is where it will be.

The standard humanist critique of science is directed at the level of programs, for example, in the impressive later work of Husserl, at mathematical idealizations, the horizons of the life-world and at the loss of contact with the subjectively meaningful acts of consciousness.[3] Accordingly, the historical European specialization in science amounts to a loss of meaning. This neither raises nor answers the question of the unity of the code, of the unity of the difference of truth and falsity. But the 'loss of meaning' (which is undoubtedly meaningful, or so many books could not have been written about it) can be explained by a differentiation of function systems, especially with the differentiation of a function system for scientific research. In this way we reach a position that suggests applying the unity of the code to the unity of the system that uses it.[4] Separate coding enables the system to be differentiated just as the system can record, standardize and subject the code to practical verification.

This information, however, remains very formal if it is not concretized in terms of the particular characteristics of the scientific code. On this score there are two arguments:

1 The code of scientific truth and falsity is directed specifically toward a communicative processing of experience, i.e., of selections that are not attributed to the communicators themselves. The intervention of personal qualities and circumstances are treated as disturbing noises and, like other 'accidents', eliminated if they do not lead to valuable discoveries of truths or falsities.[5] Historically this means that the

explanation of the causes of errors – an important part of former epistemologies – is de-anthropologized, freed from connotations like original sin, cognitive faculties and ideological blindness and placed within the structures of the scientific system itself, i.e., in its theories and methods. Errors are now mistakes, often productive mistakes, in the programming of communication and can be discovered and eliminated by coded operations.[6] The code remains universal. It applies to everything that can be experienced, even to action. But it does not serve the communication that wants to carry out, cause or even recommend an action. It does not serve to select action. From this point of view truth and falsity can also be conceived as a symbolically generalized medium of communication that serves specifically to reproduce the improbability of similar experience and that leaves the steering of action to other media.[7] Above all, it is this reduction to co-ordinated experience that differentiates the scientific code from morality and religion. After all, religion is not merely the experience of God, so it is reconciled to doubt. One has to pray for miracles.[8]

2　The scientific code of truth and falsity is directed specifically towards the *acquisition* of new scientific knowledge.[9] The mere recording, preserving and retrieving of knowledge has required little human effort ever since the invention of printing. In any event, to take care of this is not an unusual, improbable performance and would not require the code of a special communication medium. What is new has to be freed from the suspicion of being an exception or being false. This occurs through an improbable, culturally historical preference for innovation, indeed for curiosity (*curiositas*), that has to be tested and standardized itself, i.e., find its boundaries within itself (resonance) and not in its objects.[10] Everything can come under consideration. Similarly, scientific analysis does not serve to solve problems but to multiply them. It begins from problems that are solved or from problems having a chance of solution and inquires further.

Important features of modern science can be explained and allocated with the help of both these specifications, namely:

3 The way that science goes about its task rests on a *differentiation of theory and method*. Theories (research programs resulting from research programs) externalize the internal results of scientific work, i.e., apply them to the world that can be experienced by everyone. On the other hand, methods apply the code, i.e., make sure that the results can be distributed according to the values 'true' and 'false'.[11] In this case the test procedures of decision theory, game theory and statistics provide only provisional certainties.[12] But even this is a reflection of the difference of theory and method. So theories represent the openness, while methods – through the exclusion of third values – the closure of systems. Their distinction itself has an exclusively internal meaning for the system: it refers only to the system's own operations.

4 After the system worked for several centuries under these conditions it became clear where it was leading. This is something that idealization, mathematization, abstraction, etc. do not describe adequately. It concerns the *increase in the capacity of decomposition and recombination*, a new formulation of knowledge as the product of analysis and synthesis. In this case analysis is what is most important because the further decomposition of the visible world into still further decomposable molecules and atoms, into genetic structures of life or even into the sequence human/role/action/ action-components as elementary units of systems uncovers an enormous potential for recombination. At the same time, however, this potential for recombination is something with which science overburdens itself.[13] Parsons's theory of the general action-system presents this situation for the social sciences.

5 For methodological and theoretical progress this development moves the problem of the observation, description and explanation of facts with *structured complexity* – quite apart from the ways classical epistemologies proceeded – to the center of interest.[14] From the concept of the 'black box' to the radical distinction of operation and observation, the theory of self-referentially closed systems, the auto-logic of self-observation and self-description and corresponding theories

of intervention research encounters the resistance of reality forcing it to understand itself in terms of structured complexity. It follows only then that complexity is interpreted as the measure of system-ignorance and as the prerequisite for many, theoretically non-integratable descriptions.[15] This means that science has to understand itself as a system that observes observing systems (cf. above, chapter 5). At the same time, it realizes that it, too, is nothing more than an observing system that depends on its own structures. It encounters itself as a complex system that recalculates its calculations with a view towards self-provoked disturbances from the environment.[16]

6 The problem of structured complexity is not the ultimate problem (if we could solve it) that would clear the way for omniscience. Omniscience is not only factually impossible but also logically impossible (this is not to say theologically impossible!) because it would have to include itself within itself. Analogously, a system that tries to steer itself according to its own complexity is only hypercomplex. It employs some of its operations *to reduce complexity* and others to observe and describe that reduction and how it occurs. Thus it generates a second complexity that encompasses the first, plus other things. This is unavoidable if our theme of social resonance to the exposure to ecological dangers is to become the theme of scientific research. If so, then science describes all of society, including itself, as a subsystem of society. The description of description of description – *ad infinitum*, except in a self-description of science that applies to itself whatever it postulates for all systems: limited resonance according to its own frequencies and eventually according to binary coding. This is like Baron Münchhausen's story of his adventure in the swamp where he pulls himself out by his own hair – but with the possibility of seeing how the others do it!

Even the scientific system that concerns itself (with the help of this code and with its corresponding reductions) with themes like the environment, owes the system's openness and learning capacity to the closure of its autopoietic self-reproduction. Even the scientific system finds itself reduced to a self-structured resonance;

otherwise it would not be in the position to recognize information as scientifically relevant – to classify it as true or false and to accord it a self-transcending relevance through inclusion in theoretical contexts.

The possibility of recognizing these limitations resides in forcing them back into paradoxes, i.e., into undecidability. Thus, for example, all theories are directions for comparisons. The more different the thing to be compared, the more powerful the theory. To this extent science is the pursuit of the possibilities of treating as the same, things that are different. Its theories dissolve the paradox although they work with it and transform it into normal research.

Alternatively, science proceeds as the decomposition and recombination of reality. The further the decomposition is carried the more difficult, the more effective, the more 'catastrophic' the recombination. Possible examples of this include the physics of subatomic relations, modern genetics and the assumption that human behavior is the result of processes of socialization whose foundations in stratification have to be eliminated. Decomposition and recombination are carried out as a unity and just like the process of comparison, this unity is the condition of the appearance of *new* knowledge, i.e., of the acquisition of knowledge. Thereby, what is not intended has to be intended too: the increasing probability of uncontrollable recombinations.

An unavoidable concomitant of this is the classic paradox of *ceteris paribus*. The assumption of *ceteris paribus* is the condition of isolating the objects of research, but just like the presuppositions of model-formation it is a consciously false assumption.[17] Only through false assumptions can true knowledge be attained.[18]

Since the latter problem has central importance for ecological research we would like to spend some time with it. Ecological research usually speaks of *ecosystems*. This concept would be appropriate, however, only if external boundaries could be supplied. But this is not the case.[19] It does not even help to define systems by means of self-regulation instead of by boundaries.[20] Self-regulation presupposes system boundaries. If the system-premiss is abandoned and if ecological problems are not defined as internal problems of an encompassing system there still remains the possibility of working with the *Simon/Ando-theorem of near*

(almost complete) decomposability.[21] This provides a somewhat
clearer overview of the limitations. One, then, can ignore
environmental influences on the sector under research for relatively
short periods of time, or the internal relations within its
component parts for longer periods. But one cannot ignore the
influences on the growth of the component parts themselves or
their other environmental relations. *Even if* one takes complete
decomposability as the starting-point and assumes that a corresponding clarity of environmental differentiations reduces the
internal problems of the scientific system, the theorem still
demands the consideration of the objective and temporal limitations of the horizons of research. Considered methodologically,
the theorem is a correlate of the problem of the opacity of
structured complexity and, viewed theoretically, a variant of the
paradox that truths can be obtained only through the allowance
of falsities. It demonstrates the way in which abstractly formulated
problems are transformed into concrete research plans and then
create resonance within science.

These very abstract statements already indicate clearly enough
the type of resonance capacity possessed by scientific research
and its boundaries. As never before, an almost endless capacity
for resolution has revealed unbounded domains of possibility to
society. Science produces a transparent world that, wherever it
concentrates, reflects itself and transforms the transparency into
access to something new. Imagination is given wings, new kinds
of combinations are conceivable – whether as technical artefacts
or as their unwanted, perhaps catastrophic side-effects. Everything
that is possible and everything that is, is selection. But only
selection is capable of being rational. So our question is, how
can this rationality be rational if it has to select one out of an
astronomical number of other combination-possibilities? This is
not the world-picture of an idealizing and quantitative mathematics that was deplored by Husserl. Nor is it the world-picture of
technical instrumentality deplored by Habermas. Instead, this is
a world sinking both inwardly and outwardly into the void, a
world that can grasp only itself but can still change everything
graspable, a world unsuited for social orientation. If we are
permitted to extrapolate from research development so far, future
research will not reduce this picture to what is concrete and

tangible but will reproduce it everywhere it sets about its task. This makes an ultimate foundation of rationality unattainable.[22] At best the concept of rationality can be revised and adapted to this situation.

There are also plenty of counter-philosophies. One can extol moral responsibility for side-effects or emphasize the necessity of doing something. One can say that despite all else we are still doing well 'in a life-world sense'. But these are reactions of defiance: defense semantics that, if they attain communicative currency, are open to observation and then break down as a result.

Obviously, society as a whole neither wants nor is in the position to assume the scientific world-picture. This is and remains a mere implication of research. What science really exports is the consciousness of selection and technology: the consciousness of selection in reference to still-indeterminate recombination possibilities and technology as already determinate and realizable. In this way, other function systems acquire the task of sorting out what is usable and what is not. Only a fraction of what is scientifically possible is ever realized. Most is not feasible economically, legally or politically. The effects of contingency spread and, in addition to problems that they create for themselves, other systems are still not in the position to have to want what is technically possible. In this situation the ability to reject what is technically possible gains greatly in significance. It can be used against the creation of ecological risks as much as in the selection of corrective measures. It is more likely, however, that it will be practiced in the economy with a view to economic profitability, in law according to criteria of existing law and in politics for reasons of political opportunity.

13

Politics

The analysis of the function systems of the economy, law and science shows that in all three cases a code-closed, autopoietic self-reproduction is the condition of the system's openness, i.e., the condition of resonance capacity and its boundaries. The social differentiation of these function systems – and this is something they have in common – extended their horizons of possibility. But at the same time it defined more exactly where the boundaries of possible resonance in the individual function systems lay. It is no different for the political system.

Even today politics claims a special place in society, just as it always has. Traditionally – going back to Plato and Aristotle – society has been understand as a politically constituted system, as *societas civilis*. In this sense political guidance was an indispensable structural condition of a *communitas perfecta*. In the structure of social differentiation the *function* of politics was connected to a determinate, distinctive system arrangement. In the metaphor of the body, for instance, it was identified with the head or the soul. In other depictions it assumed the position of an apex or center. Even today we still expect politics to provide social integration and the solution of otherwise insoluble problems. Jean Baechler, for example, continues to view politics as the center of society. But his definition of this center is enough to cast doubt on the assumption that, 'The center of the social system is the activity that combines maximum power with maximum sensibility.'[1]

Such ascriptions were plausible for the predominant forms of structural differentiation of society. The notion of an 'apex' made sense in the context of stratified differentiation, that of a center in the context of a parallel differentiation according to center and periphery as was clearly evident in the distinction of city and countryside and in the regional differentiation of Europe and its colonies. These conditions, however, have changed. The primary structure of social differentiation at present is connected with the distinction according to function systems. Even the extent to which contemporary world-society is still differentiated according to stratification or according to center and periphery is something that results from the functioning of function systems. Political reasons have determined the regional segmentation of the political systems of world-society in terms of states despite the permanent threat of war. Economic reasons force differentiation according to center and periphery, i.e., according to highly developed and underdeveloped regions.

A political theory that does not adjust itself to the realities of functional differentiation will oscillate between overestimation and resignation concerning political possibilities and try to conduct politics with promises and disappointments – and the politics that does not admit that it is incapable of doing something is caught in this dilemma. As the force whose task it is to put things in order, politics works mainly through removing the limits to the appeal to it. It regenerates hopes and disappointments and continues to thrive because the themes in which this occurs can be changed quickly. The inclusion of ecological problems within politics may reinforce this see-saw effect because through them it becomes quite clear how much politics would have to accomplish and how little it can. So the political system is constantly tempted to try to do this through a different government, a different party, eventually through a different constitution. With this observation we are already in the midst of the analysis of the specific resonance capacity of politics and its boundaries.

In reality even the political system is differentiated by means of a special code through which it attains the closure of its own mode of operation and an openness to the environment and change of political programs. The code is commensurate with the centralization of political power in the state. Power is political only in so far as it can be used to cover collectively binding decisions, and the question concerning who is and who is not entitled to this is defined by the holding of political office. Just as with the utilization of money in a money economy, this leads – contrary to what was customary – to a considerable restriction of the semantics of politics.[2] The channeling of decisions into political offices establishes politics as difference, not as unity. The holding or non-holding of positions of political authority, especially those that regulate who has political influence in different matters and how much, is what is important. This, of course, does not mean that politics is exhausted by decisions ascribed to the state. All the activity that leads up to it is political if it seeks to influence any of the premises of these decisions – whether these are programs, forms of organization, the personal filling of offices or particular subordinate decisions. The structure of state offices serves as the political code, indeed as the unified code of *all* of politics. It defines a zero-sum principle and an either/or: positions in parliament, government and administration can only be held or not held. Politics is thereby coded according to government and opposition depending on whether political groups enjoy a parliamentary majority, occupy the presidency and other important government offices or not. As long as political movements, like the 'Greens', do not adjust to this either/or alternative, but prefer to operate in the government and in the opposition at the same time, they act without any understanding of the structural conditions of the system and the best that they can do is to make trouble.

Even in this functional domain the code's abstraction serves to mobilize and adapt programs. The holding of state offices cannot be legitimized by being held once and for all, for example, by a dynastic family. Holding such offices is contingent, a process of selecting persons and programs and is under continual examination. Political election and the formation of governments serve to bring the code and program into agreement for a certain amount of time, i.e., to hand over the government to those who personally and professionally seem to offer the guarantee of carrying out the preferred political programs. This presupposes a structural uncoupling of code and program, i.e., the possibility of opening access to other programs.

The political complexity that is attained in this way can best be understood if, for the purpose of historical comparison, the theory of 'reasons of state' (*Staatsräson*) is considered. By the year 1600 the idea of 'state government' was already detectable. But the organization of the state and the holding of offices (especially the holding of the leading offices by a prince) were not clearly distinguished in the concept of the state. Therefore 'reasons of state' were directed at a methodology of holding-on-to-the-government, while the necessity of 'reasons of state' was justified by the necessity of government itself. In other words, the concepts of domination and state were not yet separated, so that one could still say: '*L'Etat c'est moi*'. The code function of the leading offices – the fact that their occupation by one person excluded that by another – was not yet differentiated from the program function, i.e., not distinguished from the question of by whom and according to which programs is the government to be executed properly. Only the separation of these questions makes the popular will – expressed in political elections – function as a criterion, while the code value of office-holding as well as the code value of truth or the possession of property and money loses its value as a criterion. In a clearer conceptual sense one can say that the holders of governmental offices today *do not possess authority any longer*, i.e., can no longer presume to act correctly as office-holders.

According to the official presentation – which Max Weber still accepted – it appears as if political authority and the bureaucratic 'apparatus' were given over to the office-holders as the means of attaining those goals indicated by the popular will. Even the Middle Ages thought this way, except that at that time the issue was not the carrying-out of a contingent popular will but the realization of a *communitas perfecta*. The consequences of this theory, for example, the *potestas delegata non potest delegari* still concern constitutional jurists today. The goals/means scheme as well as that of delegation are, however, too simple to provide an adequate structural description. This does not obviate the fact that they can or have to be as simple as this for particular observers. We will therefore replace them with the thesis of the differentiation of coding and programming. The code of the exercise of political authority ex officio guarantees that political authority is always kept in particular hands at any time, thus that the autopoiesis of a differentiated political system continues because political power can be applied to all social power according to its own conditions. At the same time it is clear in any situation who holds the power (or who acts in its name) and who does not. Because of this difference a political opposition, bound to offering a different governmental program, can then be organized. If this condition is not met – if political opposition is not permitted (or permitted only covertly) – then it has to be purchased at the cost of the establishment of a *political stratification* of the social system. This restricts the differentiation of politics. It operates then through organization instead of coding, without the latter, thereby, gaining better conditions for an ecological politics.

Of course, the differentiation of coding and programming does not mean that there is no connection at all between them. Connections are produced when the holding of office – the positively valued code position – cannot be pursued for its own sake or for the sake of income.[4] A person has to reveal *how* he or she intends to exercise political authority and, thereby, what he or she views as correct behavior. Secondary codings have been established to bridge the difference of code and program. Ever since the French Revolution a distinction of restorative (conservative) and progressive politics has existed. Both of these can be understood differently. But it is also clear that this coding can hardly refer to the real dynamics of social change and therefore remains 'ideological'. Consequently, tendencies to replace them with the distinction of a more restrictive or expansive understanding of the state have surfaced recently.[5] In this way the binary structure of the code is copied and, at the same time, a viewpoint for selection of what is considered correct is indicated.

All this has only been a propaedeutic to our theme of the resonance capacity of the political system. It shows that even the operations of the political system follow the general course of differentiated function-systems.[6] Thus there is little sense in attributing a special social position to the political system, like a leading role or complete responsibility for the solution of ecological problems. Even the political system cannot act outside its own autopoiesis, its own code and its own programs. If this happened then such an activity would not be recognizable as politics at all. It would not offer the possibility for further connections. It would necessarily be perceived as something else, perhaps as a social movement, criminality, youthful immaturity or as a fashionable or academic curiosity. Thus even the political system is capable of resonance only within the context of its own frequencies. Otherwise it would not be a system. It is limited to executing a practicable politics alone. The conditions of practicability have to be regulated and, if necessary, changed within the system itself. It needs to be said that this is not – as in all other cases – a matter of the restrictions of natural law and of the a priori conditions of impossibility but of the consequences of autopoietic autonomy and functional differentiation.

Political resonance arises because 'public opinion', as the true sovereign, suggests the chance of re-election. Even legal offenses, scandals, etc., are redirected back into politics in this way.[7] But we must also remember that politics' sensibility to themes can be influenced by the selection of personnel too. There may be more or less ecologically open, more or less committed politicians and functionaries. This can cut across all parties throughout the bureaucracy. In view of these system sensors it can become a guide-line of political programmatics to look for the possibilities of extending the resonance capacity of the political system, as long as resonance is not mistaken for arguing [*Räsonanz*]. Otherwise an effect like the 'Green' parties results: they are completely in the right with their principles, but one simply cannot listen to them.

Fixed limitations of political resonance exist because the system's own medium of political power, which is protected by controls of physical force,[8] has little chance for application in highly complex societies. The crude application of such power is almost useless for the regulation of ecological problems because no one can be forced into any specific behavior that would improve the relation of society as a whole to its environment. Besides, political power, because it ultimately rests on the threat of physical violence, i.e., on fear, possesses the real limitation *that it can neither prohibit nor prevent anxiety.*[9] Violence can certainly be used to combat violence but anxiety cannot be used to combat anxiety. The recursive application of anxiety on itself produces peculiar situations of force within politics itself. Anxiety over what is bad as politics returns as a political factor within politics. This does not mean that the scope of resonance is extended. On the contrary, there are only a few possibilities of dealing with anxiety. It is not surprising then that the politics of the 'Greens' as long as it operates this way does not seek a rational attitude towards ecological questions but approaches the objects of its fears directly. This amounts to the politics of obstruction: no nuclear energy, no concrete *pistes*, no cutting-down of trees, no razing of houses. The limitations of the political system prevail as the restriction upon blockages, and this can be covered only by 'principles', not by the responsibility for consequences. Not least of all this means that environmental parties have their hands full trying to qualify for the responsibility of government within the context of the political code. For then they would have to meet the requirements of the universality and openness of the political system, i.e., they would have to be able to develop programmatic guide-lines for all the questions arising within the system.

A universally competent politics is restricted essentially to two measures that require intervention in the legal system and in the economic system. It can make laws under the condition of compatibility with the legal order and it can spend money under the condition that the incapacity of making further payments which results from this, can be transferred. In other words, it can use power to enforce new laws and it can use power to procure money without a return for it. Both of these possibilities increase the possibilities of order that the legal system or the economic system could produce by themselves. But both of these possibilities also require that the legal and the economic systems remain functional and are able to regenerate their specific medium. The legal system has to produce enough legal components to be able to handle law-making, and the economic system has to produce enough capitalistic components in order to be able to handle the drain of money. In both cases this is not a matter of a zero-sum game or even of a problem of the exercise of power and resistance to it but of the conditions of the ability of system performances to increase. The legal system is a system only if it can co-ordinate legal communication and legitimize it by means of recursive operations. For if such communication occurred *ad hoc* and passed away with its particular situation then it would not be recognizable as law but only as the pursuit of interests. The economic system is a system only if it can produce payments through payments – otherwise the acceptance of money would cease and the autopoiesis of the economy would come to an end. Any use of law or money as an instrument of politics encounters these barriers although it is difficult to see – and mostly only with hindsight – where they are overstepped. The political use of law and money can break through these barriers because it can use its power and threaten with force. There are many examples of this. Viewed in the long term, this occurs at the expense of the instrumentation of politics. Environmental politics will require a long-term perspective.[10]

A further restriction arises out of the national, territorial limitation of the coding of political power[11] and, for the present at least, from the lack of an effective international, legal regulation of the transformation of ecological problems into national politics. This restriction is hardly open to criticism because, since ecological politics always requires a balancing with other interests and viewpoints, it is meaningful to bind the decision about this to the mechanism that articulates political responsibility and is decided in political elections. Just as obvious are the disadvantages, since the effects of society on its environment cannot be limited regionally in many respects. The following are only some of the examples that show how much society's political resonance is restricted by territorial sovereignty: the disproportionate consumption of energy by the United States, which can afford it; final dumping of atomic waste at a nation's boundaries if possible; the avoidance of stiff legal duties by transferring production to countries which – because they have no ecological politics – offer 'location advantages', disagreement over waste gas.

But political resonance follows its own logic, not only in a spatial but also in a temporal respect, and important political restrictions reside in a wilful dealing with time. On one hand, politics has to be ready for a short-term change of political directions because of elections, and an ecologico-economic accentuation of political programs to promote voter turn-out would reinforce the effects of such a change even more. This is in marked contrast to a constantly needed, long-term ecological politics. Because of the initial situation, we will have to reckon with a continuous change between primarily economic and primarily ecological preferences, at the very least. On the other hand, politically adopted regulations are often more stable than beneficial, once they become operative. Even if their premises have long since been brought into doubt and their consequences have long since been recognized, to question what is valid a second time and to dissolve agreements again is difficult and often inadvisable politically. One never knows in advance whether something of equal value can be produced. From this point of view it is hardly advisable to pay 'political prices' for environmental goods. Misallocations in the housing and agricultural markets resulting from such policies ought to warn us.

As a result much is subjected to a change that is too fast while the rest is subjected to a change that is too slow. In any event – and this is a consolation – the individual political themes have very different temporal horizons that cannot be combined. Thereby, the temporal structure of politics agrees only to a small degree with the requirements of other systems, not to mention with changes in the ecological environment. New exhaust regulations for automobiles may be felt to be urgent and desirable in view of the 'dying-off of forests'. But as far as the business of politics is concerned this is only one argument among many. The more politics depends on co-operation and the development of consensus the greater is the probability of delays, of new unexpected initiatives and of long-since obsolete bodies of regulations.

It is proper, therefore, to ask whether a competitive democracy of the type indicated here can introduce controversial environmental themes into politics.[12] With all the willingness to be straightforward and to announce intentions, and despite the spectacular career of the theme itself, not much of this has been noticed so far. Agreement has been reached that something has to be done. But the problems have to be allowed to become urgent enough so that action can be taken without the chance of losing any votes. Hitherto problems have been created by bringing environmental themes into party rivalries themselves, i.e., creating difference in the sense that one party commits itself more than the others to long-term environmental programs even at the expense of the economy (including the loss of jobs) and campaigns with this difference. This procedure still seems too risky to the so-called 'peoples parties' who regard voter-swings of even small percentages as catastrophic. It is possible that this is changing and that ecological themes, with their own harsh demands, are supplanting the old sociopolitical ones. We will always have to reckon with limited resonance. But it is by no means settled that competitive democracy and the coding of politics have to ruin themselves with ecological themes by keeping access to the government open.

14

Religion

Theologians are included in the discussions involved with environmental problems too. Their motives and interests are not viewed with suspicion. They demonstrate argumentative competence and are undeniably of good will. But their contributions to the ecological discussion remain inadequate. To a great extent they merely repeat what has been thought and proposed elsewhere without the specific religious reference. What they have to offer are mostly commonplaces that do not raise the real problems. These are usually concealed in concrete pictures, words, admonitions and appeals. After all, they propose not to take technology, science and economic relations as the sole prevailing vehicles of domination. Instead, they believe that the latter ought to be auxiliary in the formation of a human culture within the natural condition.[1] Such things are better left unsaid. They are inadequate, and of no greater help if theologically reformulated by invoking God.

In view of this, how do things stand with the resonance capacity of the religious system? What structures make the resonance of this system possible, i.e., restrict it?

Much earlier than in the case of other systems (but therefore much more precariously too) a differentiation of coding and programming seems to have worked its way into religious questions. In very early religions the sacred remains immediately paradoxical, namely, enchantment and terror, attraction and repulsion at the same time. It is recorded in rigid (but pragmatically usable) forms of rituals, taboos, symbolically visible duplication or even mythical stories. This background provides the explanation for the success of the moral coding of religion that applies the duality of joy and anxiety to the moral code of good and bad behavior and thereby eliminates its paradox: good behavior produces the positive feeling of nearness to God, while only for morally bad behavior does He have to be feared. In this way, salvation and damnation come into view. God is a good God from Whom evil has mysteriously slipped away or serves mysterious ends. Accordingly, the cosmos acquires a valid principle from which the difference of good and evil is derived.

Before long this hierarchical form of paradox-elimination raised doubts and questions. A religious observer like Job, for instance, saw what appears necessary to the system as contingent and looked for the reason. A mere glimpse of the paradox caused problems for early religious reflection. The system-guiding reflection had to concede that religion's purpose was not moral bifurcation alone, i.e., the emphatic confirmation of the difference of good and bad behavior. The coding of religion had to cut across morality. Not that moral questions had no role to play! Quite the contrary. But because of this religion could not stake everything on the difference of good and bad behavior. As much as it may have been overgrown at times with moral cosmologies like the difference between heaven and hell, the moral classification of persons had never been its real concern. Instead this was seen, particularly in the late Middle Ages, as the work of the devil; something that God, through the intercession of Mary, had to counter. After all, the presumption of making a judgement about good and evil itself was the devil's handiwork. So the idea that morality stemmed from the devil always existed somewhere.

Consequently, the coding of religion can neither identify itself with morality nor separate itself from it (for even the devil is a power that is conditioned by religion). Its ultimate difference resides in the distinction of *immanence* and *transcendence*.[2] Transcendence is no longer understood in terms of another world or as a separate and unattainably high or low region of the world but as a kind of second meaning, i.e., as a complete, all-encompassing second version of the world where self-reference has meaning only as other-reference, complexity has meaning only as implexity (Valéry) and transcendence has meaning as what cannot be transcended.

As is characteristic for all codes, the ever-present, socially self-determining reality is duplicated by an implicit assumption. It is identified by a distinction, i.e., indicated within the context of this distinction. The unity of this difference (and not transcendence as such) is religion's code. This code has many different semantics that differentiate the religious system into different religions. Take the example of creation myths. Through the creation of the difference of heaven and earth God excluded Himself from the world. Or take the example of the institution of a difference between the sacred and the profane by which the desacralization of nature is made a condition of the *specification* of religion and so recorded irrevocably or the example of the historically confirmed adoption of the belief that Jesus is the Christ, or the late medieval/early modern binary coding of the moral disjunction, according to which the sinner can be saved through remorse and grace while the just person is lost precisely because of believing him- or herself to be just.[3] The pure difference of immanence and transcendence is enriched in many different ways and subjected to the evolution of conditions of plausibility. If society, as it appears at present, encounters new situations then this does not mean that one has to bring the coding of religion into doubt immediately. But if this is done the semantic imposition of cosmologies and theologies on the code ought to be examined critically. Accordingly, we will examine the symbolic mediations between coding and programming at the same time.

As the outline of the problem in figure 3 makes clear (cf. figure 2 in chapter 9 above) the questions here are somewhat different from those in the function systems that remain.

	code	program
unity	God	revelation
operation	immanence/ transcendence	rules of Holy Scripture

Figure 3

If it is true that the modern, functionally differentiated society forces a greater differentiation of coding and programming, then as far as religion is concerned, the problems do not reside in the reflection of the unity of the code but in the reflection of the programmatic unity of the system, its goals and the conditions of the correct attribution of its values. This is where the system comes into competition with formulas of correctness like justice, prosperity and knowledge which are already functionally differentiated. Because progress became a real possibility in this regard in the seventeenth and eighteenth centuries the programmatic of correct attribution became a problem for religion itself. The morally rooted difference of salvation and damnation, of heaven and hell, of the love and fear of God, receded. Hell completely lost its plausibility. On the other hand, two ideas gained acceptance as dogmatic articles of faith: (1) the intellectual step from 'God' to 'revealed religion'; and (2) the guidance of access to transcendence through the rules of Holy Scripture.[4] This means that de-differentiation [*Entdifferenzierung*] between the moral foundation of the code, on one hand, and forcing the awareness of contingency into the dogmatical, on the other, mutually condition and require each other. The unity of the system – the connection of coding and programming – can be secured only in a way that cannot be advocated as dogmatics any longer.

At the same time the religiously founded cosmology is discarded. Even very old traditions had abandoned a directly religious qualification of nature because this was the only way that allowed for the emergence of separate religious formations that are otherwise distinct from the world, for example, the desacralization of nature as the condition of the specification of the religious. In early modernity this desacralization of nature merely changed its reference system. It was no longer a primarily religious but became a primarily scientific or primarily economic requirement, and religion could not intervene in this process because it had to preach the same sermon.

At first an attempt was made to transfigure, glorify this difference and, thereby, to resolve the paradox of a world-order created by God that was confusing, misfortunate and inconvenient. With the rejection of these solutions, however, the order itself becomes a paradox. 'The disorder in the world is merely apparent.

And where it seems to be greatest is where the true order is still even greater. It is simply more concealed from us.'[5] The difference of immanence and transcendence is dissolved or transformed into the difference of manifest and latent. The pointing finger [*Fingerzeig*] of God (*providentia specialis*) becomes the 'invisible hand'. This allows one to exploit the fact that manifest structures are more sensitive to deviations than latent ones. This version of the code provided possibilities of deviation and sensitivities and thus saved religion.

In this way an optimistically progressive society could tranquilize itself concerning the problems it produced.[6] But in view of contemporary uncertainty about the future it will hardly suffice to imagine the grandeur of the world-order in what cannot be seen. Even the question of theodicy, of how God can permit all this – why He did not create completely impenetrable atoms, why He accepts chemical fertilizing and why He allows the heads of big business to wheel and deal – does not help either. The actual problematic is blocked by the question of the justification of God or, in any event, is not addressed. The problem resides in the *impossibility of deducing the problematic from the code*. This requires a semantics of translation for any coding, for example, a theory of theory formation (a theory of science) or the legitimization of law-making or the economic superiority of capitalism or socialism. The contemporary religious system has nothing comparable to offer in this case, and the doubts that are raised and justified by these semantics of translation produce hesitation with every attempt.

As long as this does not change – and change seems impossible – religion (or, in its name, theology), will have little to contribute to the social resonance to the exposure to environmental dangers. To be sure, it will be able to protest against deforestation, air pollution, nuclear dangers, or the excessively medical approach to the human body when these problems have acquired a certain evidence. But it will not be able to intervene with a genuinely independent form of the problem because it remains dependent on an antecedent social awareness of the latter. Wherever the assurance of meaning is required, wherever environmental experiments with far-reaching, unpredictable consequences have to be carried out is where theology admonishes and creates uncertainty – or remains silent. Strictly speaking, it has no religion to offer.[7] One almost gets the impression that religion today develops as a kind of parasite on social problem situations – parasite in the sense of Michel Serres[8] as the reintroduction of the '*tertium non datur*' into the system. In other words, religion profits from the binary structure and the exclusion of the '*tertium non datur*' in all other codes. It profits because it can provide a formula of unity for its code and thus include the excluded third possibility, the '*tertium non datur*'. But does this mean that it has to leave the programmatic of what is correct to the other social function-systems and can provide only inferior programs of its own, for example, in a fundamentalistic, concretist, miraculous, eschatological way or in the form of a 'new myth'?

Once again, the entire problematic of social resonance to social dangers is reflected here as if in a mirror. Resonance can be created only through structural restrictions, the reduction of complexity, selective coding and programming, i.e., only inadequately. For the time being at least, religion only seems to confirm this through the rejection of its own reductions and thus through the rejection of its own resonance. But if this were to continue to be its presentation of the finitude of the human world, would not everything depend, for the Christian religion, on holding fast to the certainty of being accompanied by God?

This does not lead to an environmental ethics or even to a theological exaggeration of environmentally political demands. But it is conceivable that there are ecological as well as social marginal states in which it is impossible for humanity to experience the certainty of faith and to hope for redemption. At the very least the fact that this must remain a possibility can be justified by religion.

15

Education

One might look to the education system as a source of great hope for the future because an interest in ecological questions is a priority with today's young people. Would it not be possible for the education system, especially in schools and universities, to take up this interest and develop it toward a gradual change in society's awareness of and attitude towards the environment? We no longer enjoy the eighteenth century's pedagogical optimism in education's ability to produce a complete change in humanity over two or three generations. Some say that this ecological interest is a political maneuver of one class to direct attention away from the really important questions of poverty, injustice and war.[1] Even if this is given due consideration one could still hold that schools are the place where society should prepare itself for an encounter with the environment.

But the education system is only one function system among others, and in so far as it is not concerned with the education of educators, it establishes attitudes and abilities that have to be actualized in other function systems. It works in conjunction with swings in public opinion. Perhaps it does so more slowly or more persistently, but essentially it does so without the certainty of being able to find connections with the activity of the other systems. Nevertheless, the value of education may be rated very highly if we remember how insignificant great problems appear after passing through the filter of what is normatively possible, what the costs are and how much depends on reacting to environmental consequences. No function system is always capable of producing the only correct decisions. With many practicable solutions it may be of increasing importance whether and how the ecological ramifications of their effects are heeded. But how can these problems be handled communicatively within the education system? Even this system – like all the others – is capable of resonance only with severe restrictions.

Although the primary purpose of the education system is not to process communications but to change humanity, developments parallel to other function systems are also found in it. For instance, the education system has to react to its own differentiation through a structural differentiation of coding and programming. Of course, it is more difficult to recognize how this happens than in the case of religion. The pedagogical literature is as unreliable in its way as theological literature because both begin from a programmatic orientation.

The coding of the education system is connected with its function of making social selections. This is where we find that technical bivalence that characterizes a code. A person can do well or poorly in exams, be commended or rebuked, receive good or bad grades, be promoted or not, be admitted to advanced courses or schools or not. Finally, he or she can graduate or not. Occasionally this bivalence can be broken down into scales. But even then it functions comparatively (whether temporally for the same person or socially in relation to others) as the bivalence of better or worse.

The code of the education system results from the need to develop a career, i.e., to construct a sequence of selective events which, at any time, results from an interplay of one's own selection and the selection of others and signifies the condition of possibility and structural restriction for events connected with this.[2] Only if someone is admitted to a school can he or she receive grades. Grades are important for promotion within the school. The successful completion of a program of study is important for entry into a career. Entry into a career determines advancement. In all these cases the intentional or unintentional non-fulfillment of requirements has value for the career as well, only it is negative. Those who leave this process without completing it may decide on no career at all. But they cannot avoid this reference of their behavior to careers since careers are a given possibility and since they are the standard instrument of inclusion by which persons are appointed to positions throughout many systems.

The system's programs, on the other hand, have to do with the content of what is to be learned or they describe the conditions or capacities that are expected of a person as the result of the education process. Through programs the education system is connected to social demands. On this structural level it can also be connected with ecologically important knowledge. Instead of learning when Frederick Barbarossa was born and when he died, people learn which thallium values – according to the technical guide for clean air – present the tolerance limits. After the completion of an intensified course they also learn why the limits have been set in this way, what could happen if these limits are exceeded and which arguments could be used to effect a change in these values, if necessary.

Beyond the question of studying Barbarossa or thallium, these things have a second existence according to the selection code of the education system. They can be known correctly or incorrectly or not even known at all. More depends on them than merely German history or the chances for the authorization of an increase in the output capacity of a cement works. Once introduced into the education system they have a career value – that is, whether someone has learned this well or poorly or not at all. When there is a differentiated education system, careers are unavoidable. Their coding can take up and handle any themes at all – therefore the code of social selection can and also has to outlast the change of programs, educational goals and pedagogical fantasies. But because of this it does not meet the requirements of the main concern of programs, which is to produce something correct. It merely creates a formal contingency that forces itself upon all programs. This is not a contingency of knowledge, for example, that Barbarossa did not have to be born or that the technical guide for clean air could have been formulated differently – but one that is created by the differentiation of the education system itself. In the education system's context of selection everything can be connected with everything else, provided that it has been assigned a value. Both Barbarossa and thallium enter into the grade-average that prestructures what can be connected to this.

If one takes this assumption as the point of departure it is easy to explain why, even in the education system, tendencies to separate and recombine coding and programming appear. On one hand, the pedagogical profession assumes the task of selection – the participation in examining and grading – as an unpleasant collateral business that disturbs and impedes the real task of educating. On the other, the binary code establishes itself here as in the case of the other systems, and programs are selected and used to enable the co-ordination of code values. To an extent that could only be explained after careful analysis, the student is treated as a 'trivial machine' (in distinction to a Turing machine), i.e., as a machine that has to produce the only correct answer to an input like a question or an assigned task (it makes no essential difference if a determinate scope of reaction possibilities is acceptable). Above all, this means that the momentary state of the system tested in this way, for example, whether the student is willing or attentive or is interested as such, plays no role for the instructor.[3] To the extent that the student is treated as a trivial machine the ability to integrate coding and programming is equally assured. If educators rise up in protest against this description of their premises and activity then this indicates, at the very least, that they take the difference of coding and programming more seriously than schoolmasters who take a more 'hands-on' approach. But at the same time the question becomes more urgent: how can the quality of the work of self-referential machines be fixed if Turing qualities like 'no trouble-makers' are required of the participants?

If one takes this dominance of coding as the starting-point then it becomes immediately evident that the career structure that forms the basis of selection in and out of school transfers the social pressure of problems back into the school and how it does so. This can be recognized in school as 'performance stress' or as creeping discouragement or despair over the uncertainty of finding a desirable job because of school performance. In the case of career possibilities the demands on performance and discouragement increase proportionately when it becomes clear that future prospects depend on particular performances and that these are uncertain. It is very questionable whether the determinations of attitudes that enter the system through the selection *code* can be balanced effectively on the level of programs through a better agreement of interests and curricula. Of course, this is ultimately an empirical question. But the theoretical assumption of a differentiation and recombination of coding and programming speaks against it. Within the system the curricula have to be suitable to enable a just and relevant distribution of code values, and this structural requirement is difficult to reconcile with the claim of non-trivial, self-referential machines to be taught only according to their own interests and conditions.

This necessarily brief and rough outline of only one of the many structural problems in school education[4] already demonstrates that even the education system is not free from – indeed to a great extent is overburdened by – its own structural problems and operative requirements, which leave little room for a simple reprogramming toward an increase in ecological sensibility. Of course, it is easier here than elsewhere to arrange things, whether thallium or Barbarossa. But we will have to concede realistically that social systems cannot establish a rational relation to ecological problems by this alone. One can just as well imagine that, at the same time, a further, quasi-reflexive kind of environmental pollution is created, namely, misplaced ideas, to go along with misplaced matter.

The education systems works directly only on a particular environment of the social system – the bodily and mental conditions of people. If the effects of this are felt in the social system then this environment in turn has to affect society, that is, be able to be connected with it communicatively. Thus the education system offers perhaps the best chances for an extension of intensified ecological communication – under the condition that *two* thresholds of resonance are overcome: that of the education system itself and that of all other function systems of society in which new attitudes, evaluations and sensibilities to problems are introduced through education.

It might be difficult to evaluate these possibilities realistically. In times of need and as reserves of certainty, so to speak, they might acquire greater importance than as well-coded, programmatically organized daily life.[5] Not least of all, the success of environmental politics depends on whether and to what extent coded communication can react to what participants assume as the opinions of other participants. We should also mention another consequence of education, long-term effects and value-changes: the presumption of consensus brings about social movements that do not fit within the function systems and can be experienced by them only as noise. We will come back to this in chapter 18.

16

Functional Differentiation

The preceding chapters discussed the existence of ecological problems and the ways in which they trigger resonance in the function systems of modern society. But in the analysis of particular systems the sociologist should not lose sight of the unity of society. Indeed, the comparability of function systems and certain agreements in the structures of their differentiation – we examined the differentiation of codes and programs but this is only one of many viewpoints – point to this. The unity of the entire system resides in the way it operates and the form of its differentiation. The more clearly social evolution approaches a specific kind of operation, namely, meaningful communication, and the primacy of functional differentiation *vis-à-vis* other forms of internal system-formation the more obvious its corresponding structures become. If one eliminates all anachronisms, the conceptual and theoretical means by which society describes itself in its scientific system – in this case in sociology – have to be adapted to this.

Above all, one must realize that theories of hierarchy, delegation or decentralization that begin from an apex or center are incapable of grasping contemporary society adequately. They presuppose a channelling of the communication flow that does not exist nor can even be produced. Furthermore, the attempts to describe the relation of state and economy according to the model of centralization and decentralization and then, when it is politically expedient, to praise the advantages of decentralized decision-making and to warn against its disadvantages is unrealistic. In reality, the economy is a system that is highly centralized by the money-mechanism but with a concomitant, extensive decentralization of decision-making, whereas the political system organizes the political organisation more or less centrally and handles political influences according to entirely different models, like those of social movements. These systems distinguish themselves through the way in which they try to combine and reinforce centralization and decentralization according to their respective media of communication. But their independencies cannot be understood according to the model of centralization and decentralization.

Thus it is pointless to try to conceive the unity of modern society as the organization of a network of channels of communication, steering-centers and impulse receivers. One immediately gets the impression that good intentions cannot be realized because somewhere something is directed against them[1] which frequently ends up in mythical explanations in terms of capitalism, bureaucracy or complexity. With the help of a theory of system differentiation it is evident, however, that every formation of a subsystem is nothing more than a *new expression for the unity of the whole system.*[2] Every formation of a subsystem breaks the unity of the whole system down into a specific difference of system and environment, i.e., of the subsystem and its environment within the encompassing system. Every subsystem therefore, can use such a boundary line to reflect the entire system, in its own specific way; one that leaves other possibilities of subsystem formation open. For example, a political system can interpret society as the relation of consensus and the exercise of force and then attempt to optimize its own relation to these conditions. On one hand, consensus and force are specific operations, but on the other, they are also all-encompassing formulas and horizons for social conditions and consequences that can never be made completely transparent in the political subsystem.

Every function system, together with *its* environment, reconstructs *society*. Therefore, every function system can plausibly presume to be society *for itself*, if and in so far as it is open to its *own* environment. With the closure of its own autopoiesis it serves *one* function of *the* societal system (society). With openness to environmental conditions and changes it realizes that this has to occur *in the* societal system because society cannot specialize *itself* to one function alone. This is a matter of the operationalization of a paradox. Presented as the difference of system and environment the function system is and is not society at the same time. It operates closed and open at the same time and confers exclusivity on its own claim to reality, even if only in the sense of a necessary, operative illusion. It confers bivalence upon its own code and excludes third values that lurk in the environment's opacity and the susceptibility to surprise. In this way society reproduces itself as unity and difference at the same time. Of course, this does not eliminate the paradox of *unitas multiplex.* It reappears within the system as opacities, illusions, disturbances and the need for screening-off – as transcendence in immanence, to put it in terms of the religious system's selective coding.

This systems-theoretical analysis highlights the significance and the preference of modern society for institutions like the market or democracy. Such descriptions symbolize the unity of closure and openness, of functional logic and sensibility. Of course, the market is not a real one (as it could be seen to be from the cousin's corner window)[3] and democracy no longer means that the people rule. This is a matter of a semantic coding of an ultimately paradoxical state of affairs. It explains the meaning and the illusionary components of these concepts, explains the weakness of the corresponding theories and explains why, since the beginning of the eighteenth century, a kind of self-critique has accompanied this.

Yet the unity of this order is already necessarily given by evolution, i.e., through the continual adjustment of possibilities. Evolution does not guarantee either the selection of the best of all possible worlds nor 'progress' in any sense. At first evolutionary selection produces a very improbable, highly complex order. It transforms an improbable order into a probable (functional) one. This is exactly what concepts like negentropy or complexity intend. But it does not mean that the improbability disappears or is inactualized as prehistory. It is co-transformed and *'aufgehoben'* in Hegel's famous sense. It remains a structurally precipitated risk that cannot be negated.

Stratified societies already had to deal with problematical

consequences of their own structural decisions. These were expressed, for example, as the constant conflict between inherited honors and distinctions and new ones, as the unfulfillable obligation to prescribe a class-specific endogamy and not least of all as the conflicts that result from centralizing the control of access to scarce resources, above all of the ownership of land. Compared to modern society these are relatively harmless problems for which historically stable solutions were found in many cases. The transition to primarily functional differentiation leads to a completely different constellation with higher risks and more intensified problems resulting from structural achievements. Society's self-exposure to ecological dangers is therefore not a completely new problem. But it is a problem that, today, is coming dramatically to the fore.

With functional differentiation the principle of elastic adaptation through processes of substitution becomes the principle of the specification of subsystems. Its consequence is that, more than ever before, functional equivalents can be projected and actualized *but only in the context of the subsystems and their coding*. Extreme elasticity is purchased at the cost of the peculiar rigidity of its contextual conditions. Everything appears as contingent. But the realization of other possibilities is bound to specific system references. Every binary code claims universal validity, but only for its own perspective. Everything, for example, can be either true or false, but only true or false according to the specific theoretical programs of the scientific system. Above all, this means that no function system can step in for any other. None can replace or even relieve any other. Politics cannot be substituted for the economy, nor the economy for science, nor science for law or religion nor religion for politics, etc., in any conceivable intersystem relations.

Of course, this structural barrier does not exclude corresponding attempts. But they must be purchased at the price of dedifferentiation (*Entdifferenzierung*), i.e., with the surrender of the advantages of functional differentiation. This can be seen clearly in socialism's experiments with the politization of the productive sector of the economy or even in tendencies towards the 'Islamization' of politics, the economy and law. Moreover, these are carried out only partially. For example, they do not touch on money (but, at best, the purely economic calculation of capital investment and prices) and are arrested by an immune reaction of the system of the world society.

The structurally imposed non-substitutability of function systems does not exclude interdependencies of every kind. A flowering economy is also a political blessing – and vice versa. This does not mean that the economy could fulfill a political function, namely, to produce collectively binding decisions (to whose profit?). Instead, the non-substitutability of functions (i.e., the regulation of substitution by functions) is compensated by increasing interdependencies. Precisely because function systems cannot replace one another they support and burden one another reciprocally. It is their irreplaceability that imposes the continual displacement of problems from one system into another. The result is a simultaneous intensification of independencies and interdependencies (dependencies) whose operative and structural balance inflates the individual systems with an immense uncontrollable complexity.

This same state of affairs can be characterized as a progressive resolution and reorganization of the structural redundancies of society. The certainties that lay in multifunctional mechanisms and that specified systems for different functions and programmed them to 'not only/but also' were abandoned. This is shown very clearly by the reduction of the social relevance of the family and morality. Instead, new redundancies were created that rested on the differentiation of functional perspectives and *'ceteris paribus'* clauses. But this does not safeguard the interdependencies between the function systems and the social effects of the change of one for the other. Time, then, becomes relevant: the consequences result only after a certain amount of time and then they have to be handled with new means that are, once again, specific to the system. This is accomplished without being able to go back to the initiating causes. Complexity is temporalized[4] and so are the ideas of certainty. The future becomes laden with hopes and fears, in any event, with the expectation that it will be different. The transformation of results into problems is accelerated, and structural precautions (for example, for sufficient liquidity or for invariably functional legislation) are established so that such a reproblematization of the solution is always possible.

The rejection of substitutability has to be understood essentially as the rejection of redundancy, i.e., as the rejection of multiple safeguarding. As we know, the rejection of redundancy restricts the system's possibilities of learning from disturbances and environmental 'noises'.[5] This implies that a functionally differentiated system cannot adapt itself to environmental changes as well as systems that are constructed more simply although it increasingly initiates concomitant environmental changes. But this is only part of the truth. For, through abstract coding and the functional specification of subsystems, functional differentiation makes a large measure of sensibility and learning possible on this level. This state of affairs becomes quite complicated when many system levels have to be kept in view at the same time. Society's rejection of redundancy is compensated on the level of subsystems, and the problem is that this is the only place that this can occur. Family households, moralities and religious cosmologies are replaced by an arrangement in which highly organized capacities for substitution and recuperation remain bound to specific functions that operate at the cost of ignoring other functions. Because of this the consequences of adaptive changes are situated within a complex net of dependencies and independencies. In part, they lead to unforeseen extensions, in part they are absorbed. In such cases simple estimations and simple comparisons of the efficiency of different social formations are insufficient and inadvisable.

A further consequence of functional differentiation resides in the intensification of apparent contingencies on the structural level of all function systems. Examples of this are the replacement of natural by positive law, the democratic change of governments, the still merely hypothetical character of the validity of theories, the possibility of the free choice of a spouse and not least of all everything that is experienced as 'a market decision' (with whoever or whatever may decide) and is subjected to criticism. The result is that much of what was previously experienced as nature is presented as a decision and needs justification. Thus a need arises for new 'inviolate levels' (Hofstadter), for a more rational and justifiable a priori or, finally, for 'values'.[6] Evidently, the strangely non-binding compulsion of values correlates to a widespread discontent with contingencies as much as to the fact that decisions become more exposed to criticism through structural critique and statistical analyses than facts. Indeed, even if we cannot determine that someone has decided (for example, about the number of deaths from accidents or about the increase of the rate of unemployment) decisions are still necessary to redress these unsatisfactory conditions. To require decisions means to appeal to values, explicitly or implicitly. Consequently, structural contingency generates an order of values without considering the possibilities of concretely causing effects, i.e., without considering the attainability of the corresponding conditions.

It is probable that ecological communication will intensify this inflation of values even more. For if society has to ascribe environmental changes to itself then it is quite natural to reduce them to decisions that would have to be corrected: decisions about emissions quotas, total consumption amounts, new technologies whose consequences are still unknown, etc. We already noted in chapter 3 that such ascriptions are based on simplifying, illuminating and obscuring causal attributions. This does not prevent them from being carried out and communicated, but, if nothing else, it permits values to surface.[7]

At first, one might think that the value of clean air and water, trees and animals could be placed alongside the values of freedom and equality, and since this is only a matter of lists we could include pandas, Tamils, women, etc. But viewed essentially and in the long run this would be much too simple an answer. The problematic of the inflation of values as a symbolically generalized medium of communication – an idea of Parsons's[8] – results from its influence on society's observation and description of itself.

Actually the descriptions of society are steered by the problems that result from structural decisions and, therefore, they have a tendency to evoke values and see 'crises'. Contrary to the mature phase of bourgeois–socialist theories in the first two-thirds of the nineteenth century disadvantages are deferred for a time, are read off in values and are understood as the indefinite obligation to act. In any event, they are no longer understood as digressions of the spirit or matter on the way to perfection. Instead, they are the inescapable result of evolution. According to the theory proposed here, they are consequences of the principle of system differentiation and of its making probable what is improbable.

Moreover, the critical self-observation and description that constantly accompanies society has to renounce moral judgements or end up getting lost in a factional morass.[9] Instead, a new kind of schematism, namely, manifest or latent (conscious or unconscious, intentional or unintentional) takes its place. Only manifest functions can be used to differentiate and specify because only these can be transformed into points of comparison or goal-formulas. This means that the critique is formed as a scheme of difference that also illuminates the other side, the counterpart. Straightforward striving toward a goal is viewed as naive. This even undermines the straightforward intention of enlightenment.[10] A mirror is, as it were, held up to society, assuming that it cannot look through it because that which is latent can fulfill its function only latently. This is the way sociology, too, pursues 'enlightenment' [*Aufklärung*] and explains its ineffectuality in the same process.[11] In this sense ideology, the unconscious, latent structures and functions and unintended side-effects all become themes without a clarification of the status of this shadow world – note especially the reversal of Platonic metaphysics. One can therefore use this distinction only to discover that society enlightens itself about itself.

The problem of reintroducing the unity of society within society or even of expressing it in it is extended to the forms of the system's critical self-description. Equally symptomatic are all attempts at judging and condemning society from the exalted standpoint of the subject, i.e., *ab extra*. This signifies nothing more than placing the unity of society in a principle outside itself.[12] A systems-theoretical analysis of such attempts, however, enjoys the advantage of being able to retrace this problematic back to the structure of modern society (which changes nothing about the fact that this must occur in society).

Essentially, every attempt within the system to make the unity of the system the object of a system operation encounters a paradox because this operation must exclude and include itself. As long as society was differentiated according to center/periphery or rank, positions could be established where it was possible, as it never has been since, to represent the system's unity, i.e., in the center or at the apex of the hierarchy. The transition to functional differentiation destroys this possibility when it leaves it to the many function systems to represent the unity of society through their respective subsystem/environment differences and exposes them in this respect to competition among themselves while there is no superordinate standpoint of representation for them all. To be sure, one can observe and describe this too. But the unity of society is nothing more than this difference of function systems. It is nothing more than their reciprocal autonomy and non-substitutability; nothing more than the transformation of this structure into a togetherness of inflated independence and dependence. In other words, it is the resulting complexity, which is highly improbable evolutionarily.

17

Restriction and Amplification
Too Little and Too Much
Resonance

Detailed analyses of society's resonance to dangers from its environment have to be connected with the renouncement of redundancy that resides in the non-substitutability of function systems. This forces a channelling of all disturbances into one or several of these function systems. Whatever appears as environmental pollution can be treated effectively only in accordance with one code or another. Of course, this does not exclude the possibility that an eclipse or an earthquake can prove upsetting in other, unspecified ways. As already mentioned in reference to general systems-theoretical and particularly in reference to biological investigations, the renouncement of redundancy leads to a restriction of the capacity of reacting to disturbances (noise). On the other hand, structural restriction is also a way of increasing resonance capacity, as is quite clear from the case of the organism. Eyes and ears, nervous systems and immune systems that, for their part, are capable of resonance only within narrow but evolutionarily proven frequency ranges, develop only through a considerable amount of rejection. These reductions can then be balanced only through the organized capacity to learn.

This is the path modern society seems to have taken with its choice of functional differentiation. It would be pointless to ask whether there might have been other possibilities. It would be equally pointless to ask whether we could transport ourselves into a 'post-modern period' or are even in the process of completing this transition. The actual circumstances offer no kind of support for this. Instead, such assumptions are only premature inferences from simplistic theories. The only meaningful question must be whether we can use the renouncement of redundancy that inheres in functional specification better than before. After the logic of functional differentiation and its resulting problems are discovered we can determine how the restriction of resonance operates through the coding of the individual function-systems. On one hand, this excludes all point-for-point relevance and renounces the establishment of 'requisite variety'. Coding causes a sharp reduction for every function system. When environmental changes trigger resonance in self-referential function systems this is an exception. Only in these cases do environmental changes disturb and change the conditions of the continual reproduction of system-specific communication.

On the other hand, this reduction is the condition for noticing and processing environmental changes within the system as such. Coding is the condition that permits environmental events to appear as information in the system, i.e., to be interpreted in reference to something, and it causes this in a way that allows consequences to follow within the system. Of course, these chains of consequences are mediated by further conditions: by the system programs, for example, theories, laws, investments or party-political alignments. If ecological problems pass through this double-filter of coding and programming they acquire internal relevance for the system and possibly far-reaching attention – but only in this way!

Despite all this, there is no guarantee that society as a whole will protect against or can even contend with exposure to ecological dangers in every case. On the contrary, society can react *only* in exceptional cases. This implies that it brings *too little resonance* into play for the exposure to ecological dangers. At present this inference matches what public opinion suspects. So social communication is alarmed and stimulated to more activity without, of course, being able to translate this requirement into the language of the function systems. But this is only half the problem. The other half is more difficult to discern and at present overlooked to a great extent. There can also be *too much resonance* and the system can burst apart from internal demands without being destroyed from outside.

The problem of resonance capacity, since it refers to a *differentiated* system, does not reside *simply in one dimension* where 'too little' and 'too much' could be balanced against each other. Instead, two system boundaries have to be distinguished: society's external and internal boundaries. By means of external boundaries society screens off its own autopoiesis, i.e., communication, from the enormous complexity of non-communicative states of affairs. On this, the level of its own operations there is neither input nor output. Society cannot communicate *with* but only *about* its environment according to its capacity for information processing. Society itself, thereby, regulates what constitutes information for society. But it can also be influenced in the selection and ordering of communication by irritation and disturbances, particularly by the conscious processes of the participants.

Entirely different circumstances pertain at the internal boundaries of the system. This is where there are communicative interdependencies. The aggregate data of the economic system – growth-rates, unemployment numbers, inflationary and deflationary developments – influence the political system. Even if the function systems are differentiated according to their own autopoiesis, codes and programs, they can be disturbed by communication in a way that is entirely different from the way society itself relates to its environment. It is therefore highly probable that the turbulences of one function system are transferred to others even if, and because, each proceeds according to its own specific code. For example, the economy is at the mercy of scientific discoveries and technological innovations as soon as these find economic use. The same is true *mutatis mutandis* for the relation of politics and law, for science and medicine and for numerous other cases. There is no superordinate authority that would provide for measure and proportionality here. Through resonance small changes in one system can trigger great changes in another. Payments of money to a politician that play no role in the economic process – measured by the hundreds of billions of dollars that are transferred back and forth daily – can become a political scandal. Theoretically insignificant scientific discoveries can have agonizing medical results. Legal decisions that hardly have any effect on other decisions in the legal system itself can form road-blocks for entire political spheres. If law, for example, brings the pharmaceutical industry and physicians under the threat of liability to supply information and to establish precautionary standards then this is something that can have medical and also economic consequences that are entirely unrelated to what is legally important and might not even comprise part of the legal decision itself: the effects of anxiety, uncertainties, increase of the necessity of experiments on animals, the rise of costs or even the increase of the routine use of experimental apparatuses. In all these circumstances there is no supervening reason because every system can create resonance only with its own code. This is something that follows almost automatically if information triggers code-specific operations.

Furthermore, within their own domains, the function systems depend on *other* functions being fulfilled *elsewhere*. Particular deficits in performance can amount to unmanageable changes of the social environment of other systems and thereby produce disproportionate consequences. In this way, far-reaching economic or political consequences may ensue if the legal system, for whatever internal reasons, is incapable of developing rules for the right to conduct labor disputes that enable the participants to foresee the legal consequences of their behavior. The principle of 'proportion' [*Verhältnissmässigkeit*] within the legal system may then have disproportionate consequences in relation to the other systems. For similar reasons politically justified intervention has, on more than one occasion, ruined entire economic domains or made them dependent on constant political attention. The relationship of politics and law shows the same thing in a spectacular way. On the other hand, the stability of governments depends on whether the economic production of wealth rises or falls, i.e., on developments that are to a great extent out of their control and which often do not work out as well or poorly for the economy as they do for the political system.

For these reasons a much greater amount of resonance is more likely to occur within society than to result from its relation to the external environment. Function systems are differentiated, coded and programmed for functionally specific high output. They constantly scour their socially internal environment for impulses and pick up what is offered to them. They are endogenously restless and very sensitive. Their structural improbability – the incorporated risk – can be released very easily. If these systems describe themselves as an 'equilibrium' then this also means that they have made instability their principle of stability – or to formulate this in the terms of an eighteenth century author: a kernel of corn thrown onto one of the scales of a balance is enough to upset the system.[1]

The autonomy of the autopoiesis of the particular function systems and the rejection of reciprocal substitutability are the basis for the possibility of disproportionate reactions because every system that is *solely and completely* responsible for its *own* function regulates the conditions of the oscillation of resonance independently. But at the same time it cannot control the environmental occasions that trigger this. There seems to be no general rule for such situations. They do not show up in all of the intersystem relations, but in some of them more often than in others. They fail to appear where they might be expected, for example, in the relation between recent German divorce laws and the willingness to get married – and they appear surprisingly where they are unexpected. They can be observed and analysed. They can be understood and described as a structural property of modern society, but they cannot be anticipated.

In view of all this, the special position of the political system of modern society in relation to the latter's ecological problems would have to be examined more closely and become the subject of empirical investigations. The political system's own method resides in the production of collectively binding decisions. These have no directly ecological, but only socially internal effects. They facilitate and suppress communication, but at the same time this system reacts very sensitively to itself, and although it cannot regulate other systems via binding decisions it can influence them. Under these circumstances it is highly probable that politics becomes the place to start the business of addressing ecological issues. Precisely because the system cannot do anything here immediately it becomes increasingly probable that this is where communication about ecological themes will find a home and expand. There is nothing within the system to prevent this. Viewed from a purely political point of view, there is nothing that would correspond to legal, economic and scientific restraints and would forthwith reduce communication to what is possible. The system enables and promotes loose talk. As we can read in the newspapers, nothing prevents a politician from demanding, proposing or promising the ecological adjustment of the economy. But a politician is not obliged to think and act economically, and so does not operate at all within the very system that his or her demand will ultimately bring to ruin.

Political communication is always concerned exclusively with which political programs will or will not help the government and the opposition to take over from the other. This is its code. So communication cannot be taken to an illusory extreme because this would be observed and judged by the voters. It must therefore promote plausible decisions about law and money. It can do this as long as the legal and economic systems provide the political system with some room to maneuver. But at the same time – as we can see clearly – it relies on the effectiveness of illusions and carries out its business in this way.

Under these circumstances we must realize that politics is used as a launching-pad and transmission system for ecological desiderata when and wherever these enter the consciousness of individuals and social communication. Then the political system may function as a kind of continuous-flow heater. But this only increases the probability that, on the occasion of the exposure to ecological dangers, a socially internal intensification of resonance will result that combines politically convenient and acceptable solutions with functional disturbances in other systems. Such an oscillation of resonance will probably have destructive consequences within an evolutionarily highly improbable social system, therefore any claim to political rationality would have to include reactions to the effects of politics in its calculations.

18

Representation and Self-observation
The 'New Social Movements'

It contradicts every principle of social differentiation to re-establish the totality of the system within the system. The whole cannot be a part of the whole at the same time. Any attempt of this kind would merely create a difference in the system: the difference of that part which represents the totality of the system within the system *vis-à-vis* all the other parts. The presentation of unity is a production of difference, thus the intention itself is already paradoxical, self-contradictory.

Traditional societies however, have been able to live with this paradox. The form of their internal differentiation was able to accommodate it sufficiently. As long as these societies evolved toward advanced culture they were differentiated either hierarchically or according to center and periphery, mostly through a combination of both. At least then there was no competition with respect to the subsystem that represented the whole within the whole. Only the apex of the hierarchy, only the highest stratum, or only the center, the city and the urban form of politically civil life came into consideration.[1] Not until the Middle Ages did discrepancies arise between aristocratic and urban life that initiated a transformation process that led to functional differentiation.

Besides, the traditional advanced cultures could always resort to a religious justification of representation. This was not only a mode of (always doubtful) legitimization for them but the representation of the whole within the whole could also make use of the code of religion to articulate the paradoxically created difference. The difference that results from the attempt to reintroduce the whole within the whole and to regulate the system from an internal standpoint presents itself as the difference of what is determined in this way and a transcendent indeterminacy. 'The split that cuts across human space is the generator of an ultimate indeterminacy.'[2] It was precisely the inaccessibility of this indeterminacy that was used as a difference to create order in this world. In the eighteenth century it became evident, in view of the new, highly complex function systems, that all 'natural' representation rested on presumption and that religion is misused if it persists in concealing this presumption. This clarification reflects the transition from stratificatory to functional differentiation. In the new order there are no natural primacies, no privileged positions within the whole system and therefore no position *in the* system which could establish the unity *of the* system in relation to its environment.

Even if we concede all this, society nevertheless cannot do without self-observation. Not all communication falls within the context of the primary subsystems, for even if this were the case, it could become the subject of communication in the next moment. All order, every form of differentiation that is realized in society, can also be observed and described in society. Every binary code that excludes third possibilities for the coded operations also makes it possible to introduce these third possibilities. All reduction of complexity preserves complexity. Other, unrealized selections are 'potentialized', transformed into what is merely conceivable and preserved for communicative reactivation.

Formally, the concept of observation indicates an operation that designates other operations within the context of the distinction of 'this instead of that'. Observation refers to operations or complexes of operations (systems). It applies a distinction as its own meaning-schema (for example, earlier/later, useful/detrimental, fast/slow, system/environment) that is not necessary and often inaccessible for the autopoiesis of the operations itself. It brings much richer meaning-possibilities into play to reduce them through selective designation. Therefore, on the level of operations, autopoietic operation and observation have to be distinguished by a scientific observer who applies this theory, i.e., this distinction of operation and observation.[4] On the system level, however, at least for meaningfully operating, conscious or social systems, one has to assume that these cannot eliminate self-observation. Therefore, as soon as forms of differentiation together with their consequences are revealed, it is probable that they are observed and described in society itself. The only question is, with the help of what distinctions?

Social self-observation has to be distinguished from other-observation within society. Of course, system differentiation also makes possible the observation of one subsystem by another. The peasants observe the nobility, the nomads the settlers, the politicians the economy, the law-makers politics, etc. One system's distinctions – especially its binary codes – are thereby applied to other systems that do not use these distinctions to observe. This is nothing more than the normal technique of reduction in the relation between system and environment, transferred to the internal system level of subsystems and their environment. Social self-observation occurs only on the condition that observation does not distance itself from its object but co-intends itself.

The dogma of original sin was a schema of self-observation unequaled and unsurpassed historically. It led, if not on the psychological then at least on the communicative level, to moral self-condemnation and therewith to a mitigation of moral criticism. No one, for example, could recognize sin in another's act of going to confession. Everyone had to do this. All classes, even the clergy, were subject to this principle. It was designed to be class-neutral and at the same time made it possible to work out a class-specific catalogue of sins and dangers to salvation. This was discussed as 'pollution' or 'hereditary pollution' of souls. Only because this schema was increasingly undermined by the personal attribution of guilt and the impossibility of knowing an individual's state of grace could a religious moralism flourish whose secular after-effects are still felt today. But a modern functional equivalent for original sin is not on the horizon.

Since the nineteenth century the self-observation of society (as an observation of this observation can determine) has been connected with prominent consequences of function systems – above all, with the resulting possibilities of the 'revolution' of the political system and with the consequences of a money economy (capitalism). This requires the causal attribution of these consequences and leads, according to all the teachings of attribution theory, to an irremediable conflict of attribution.[5] The self-observation of society becomes ideological, i.e., in the application of causal attributions dependent on valuations and partisanship.

The semantics resulting from society's observation and description of itself has, in the meantime, become historical and is still represented with the labels 'neo-' and 'post-'. In view of such rapid changes the self-description of society is temporalized, indeed contracted into a mere 'definition of the situation'. Designations like 'industry', 'capitalism', 'modernity' are retained but used only to characterize a historical difference. Of course, this makes them more appropriate, but the question whether they are valid or not becomes vacuous when one maintains that they are no longer valid. This amounts to a weakly concealed paradox: society is that which it is not and temporalization serves to awaken the illusion that, despite this, it really means something.

With such standards of orientation 'theory deficits' should come as no surprise. New kinds of social movements and social protests look for new forms of articulation. Society is still viewed as the cause and the object of protest, but the themes of protest have clearly changed to ecological ones.[6] Even the theme of peace is viewed from this perspective in so far as it concerns armaments. The politics of the military ought to be prevented from using nature against humanity. In any event, the main concern is no longer the legal theme of peace through authority.

The largely unorganized resonance of such themes is reducible to the appearance of doubts in the experiences of daily life about the meaning of nearly all function systems. The horizons of what is possible have expanded so radically that every non-realization must have causes in society. Intentions have unintended results while good intentions have bad side-effects. Rationality appears increasingly as perverted and in communication inevitably encounters mistrust and rejection. There is no doubt that such a mood is fed by experience, but at the same time it is difficult to fix the attribution to causes in a precise way, therefore the initial situation generating the ideology, no longer exists although ideologies are functionally indispensable wherever an assessment of values matters.

Perhaps these tendencies can best be characterized in reference to the binary codes of the major function systems. Attempts are made through human understanding to avoid the tension between having and not having and to mollify the sharpness of the difference between legal and illegal. Attempts are made to bring the environment to bear against the functionally rational codings of society. All in all, attempts are made to assume the position *of* the excluded third (code value) and then to live in society *as* the included excluded third (code value): as a parasite.[7]

As far as sociological observation of this observation is concerned it becomes attractive to imagine that this is ultimately *a protest against functional differentiation and its effects.* But even if this is the general name for a new kind of self-*observation* of society, a corresponding self-*description* capable of fixing the results is still lacking. The new social movements have no theory. They are also incapable of controlling the distinctions in which they record their observations. A simple, concrete fixing of goals and postulates, a corresponding distinction of adherents and opponents and a corresponding moral evaluation therefore predominates. Implicit in this is the idea that a person has to be able to live as he or she wants to even if in reduced circumstances and foregoing luxuries (whereby these sacrifices are to a great extent consensual because they are in agreement with the life-world experiences of everyone anyway).

Even the so-called early socialists, not to mention Marx, had much more to offer theoretically, but through a radical reduction of society to economy they created for themselves an initial basis that was much too narrow (and theoretically incapable of realization). It would be unfair to measure this by contemporary standards. Nevertheless, one can still see (from the position of second order observation, of course), how this kind of social self-observation operates with an inadequate semantics.

The most important effect is that observation cannot include and reconstruct within its own concept that which it protests against. This remains visible only as a resistance that results from rejected valuations. Last but not least, this contains a renunciation of its own semantics and structural stability that can be attained – here again Marx is the great paradigm – through a theoretical construction encompassing action and resistance. The blasé moral self-righteousness observed in the 'Green' movement conceals only superficially the ever-possible relapse into resignation.

The problem seems to be that one has to recognize the dominant social structure – whether seen as 'capitalism' or 'functional differentiation' – to assume a position against it. Today this is not as easy as in the nineteenth century because the hope for a historical resolution of the difference, i.e., the hope for revolution, no longer obtains. A functional equivalent for the theoretical construct 'dialectics/revolution' is not in sight and therefore it is not clear what function a critical self-observation of society within society could fulfill. Nothing more than a resigned comment on decline in the manner of Adorno or Gehlen is detectable, and such positions offer hardly anything to hold on to.

Fortunately, there is an essential connection between modern society's (semantic) deficits of self-description and the (structural) system-form 'social movement'. As a position for the description of society within society the movement places itself in difference to society. It seeks to affect society from within society as if it occurred from outside. This paradox creates the instability of the observation position and the dynamics of the social movement makes allowances for this without realizing it. This may very well lead to changes, to semantic or structural results that in one way or another come to terms with the facts. Like the 'Reds' (liberal theologians, according to Harnack) the 'Greens' will also lose color as soon as they assume office and find themselves confronted with all the red tape. This outlook may serve to appease 'conservative' observers, but it ought not to conceal the fact that the real problem resides in the question of whether modern society is too dependent for self-description on the entirely inadequate basis of social movements.

19

Anxiety, Morality and Theory

On the level of its autopoietic operations modern society is bound to a functional differentiation, coding and programming, which in turn can be seen and judged critically on the level of self-observation. But since this society cannot represent itself within itself it cannot bestow the normative sense for which one could at least assume, if not attain, a thorough-going consensus. Therefore, a self-observation cannot, as with the prophets who enjoyed a privileged position, remind us of what is essential and lament its downfall. Instead, anxiety is chosen as a theme,[1] to replace the difference of norm and deviation. This leads to a new style of morality based on a common interest in the alleviation of anxiety, not on norms for which all that matters is the avoidance (or regulation, or regret) of deviation so that people can live without anxiety.

This structure is closely connected with the differentiation of coding and programming discussed in detail in the preceding chapters. Throughout society that which is right or correct is assigned to function systems and articulated exclusively as interchangeable programs that organize the assignment of the values of the respective codes. Anxiety, then, becomes the functional equivalent for the bestowal of normative sense; and a valid one since anxiety (as opposed to fear) cannot be regulated away by any of the function systems. Panic cannot be prohibited – as Shaftesbury had already recognized in his day.[2] Anxiety cannot be regulated legally nor contradicted scientifically. Attempts at a scientific clarification[3] of the complicated structure of the problems of risk and certainty only supply anxiety with new nourishment and arguments.[4] One can attempt to buy off or to compensate anxiety with money. But whoever does so only indicates that he or she had no anxiety: the commodity disintegrates at the conclusion of the agreement. Religion, too, would only devalue itself if it tried to present itself as a means of removing anxiety. As its history shows, religion merely transfers the anxiety to other domains of meaning.

So anxiety cannot be controlled from the function systems, but is protected against all of them. As a matter of fact, better functional performance can go hand in hand with increased anxiety without being able to remove it.[5]

In this way anxiety does not have to be present at all. The communication of anxiety is always authentic since a person can attest to personal suffering of anxiety without fear of contradiction. This makes anxiety an attractive theme for the kind of communication that wants to observe function systems and describe them from outside themselves but still from within society. Anxiety resists any kind of critique of pure reason. It is the modern apriorism – not empirical but transcendental; the principle that never fails when all other principles do. It is an 'Eigen-behavior' that survives all recursive tests;[6] one that seems to have a great political and moral future. We are fortunate, then, that the rhetoric of anxiety is not in the position to create anxiety in fact,[7] even if it remains a disturbing factor within society.

In many respects the rhetoric of anxiety is not a new phenomenon. Its political use, directed against internal and external enemies, has been known for a long time,[8] but the new ecological themes have changed its direction and displaced the difference of the friend/foe schema into a system/environment perspective. To the extent that war amounts to an intentionally produced ecological catastrophe the anxiety over war suppresses the anxiety over the enemy. The old social differentiations of the national, class and ideological type that used to lead to war lose their power to convince and are replaced by reduced tendencies toward regional or cultural ethnogeneses. In reference to the 'real' problems of our time new social solidarities are extolled and postulated morally with great emphasis.

Above all, the new themes of anxiety enjoy a new property: a person need have no anxiety over showing anxiety. They are therefore capable of being extended. No one who shows anxiety in 'crises' or over ecological developments, the consequences of technology and the like, appears in a negative light, for there is no individual capacity with which the danger could be met. This enables opinion polls to record the increase of anxiety and direct its results back into public communication. In this regard talk of the age of 'unconcealed anxiety' has gained currency[9] and anxiety can claim to be universal: *volonté générale*.

Furthermore, it is remarkable that this type of anxiety inherited not only non-contradictability but also its paradoxical constitution from the rhetoric of principles. In other words, the attempt to ease anxiety increases it.[10] Even official policies and the constant concern with the improvement of circumstances can increase anxiety, for example, increasingly detailed packaging labels on medicines or intensive research and reports in the chemistry of the health sciences lead to the impression that nothing is harmless and everything is contaminated. The psychological basis for this paradoxical effect seems to be that highly improbable risks are overestimated and that risks to which a person is unavoidably exposed are thought greater than those to which a person exposes oneself intentionally.[11] Above all, one has to realize that communication about anxiety makes communication about anxiety possible and in this sense is self-inducive. One can always take a position with respect to anxiety. Some speak of 'hysteria' while others of 'appeasement' – and both sides are supposedly right.

All this reveals that not only socially dominant, functionally related communication but also anxiety-related communication is a principle of resonance that emphasizes certain things and de-emphasizes others. This difference is only intensified by a public rhetoric of anxiety. This rhetoric assumes the task of effecting anxiety (which does not show up by itself, is not self-evident). To do this it must proceed selectively.[12] So, today, there is great anxiety over nuclear dangers,[13] but hardly any against medically induced diseases, and there is an absence, at least in public communication, of anxiety over the anxiety of others. The rhetoric of anxiety is selective because it emphasizes the development of what is worse and says nothing about any progress that has been made, for example, in the medicine of the health sciences.[14] Through public rhetoric anxiety has become stylized as the principle of self-observation. Whoever suffers anxiety is morally in the right, particularly if it is anxiety on behalf of others and this can be assigned to a recognized non-pathological type.

Despite these clearly semantic contours no (sub)system can be differentiated out for the management of anxiety. Even in view of the tactful, considerate, understanding treatment of this syndrome it remains to be seen to what extent this is merely a matter of 'pluralistic ignorance'.[15] If no one really feared a radioactive contamination of the water-supply but everyone assumed that others feared this and, consequently, accepted this argument, then how could anyone detect that the anxiety was not contrived?

Of course, the social problem lies less in the psychical reality of anxiety than in its communicative actuality. If anxiety is communicated and is not contested in the communicative process it acquires a moral existence. It becomes a duty to worry and a right to expect participation in fears and to require standards for defense against danger. Therefore, those who worry about ecological matters do not, like Noah once did, equip just their own ark with the necessary material for later evolution. They become warners – with all the risks that this implies.[16] In this way anxiety infuses ecological communication with morality and controversies become impossible to make decisions on because of their polemical origin. Only the future can indicate whether the anxiety had been justified – but the future is constituted anew in every present.

Against a morality that propagates anxiety-related distinctions theoretical analyses are in a difficult position. Anxiety, since it transforms the uncertainty of the situation into the certainty of anxiety, is a self-justified principle that needs no theoretical foundation. It can, and indeed quite rightly does, ascribe theories to the function system of science and distinguish, accordingly, whether they sympathize with the anxiety or not. As a result of the long-standing apriorism of reason, the position from which an anxiety-based rhetoric and morality makes its observations enjoys an unassailable self-certainty.

On the other hand it is difficult to see how from this vantage-point the relation of society to its environment can be drastically improved. Even anxiety limits and intensifies resonance. In fact, it is more likely to stop the effects of society on its environment, but it has to pay for this by risking unforeseeable internal reactions that again produce anxiety.

If observing means distinguishing and designating one should begin from the distinguishing capacity, i.e., to distinguish distinctions. The systems-theoretical distinction of system and environment, treated consistently, aims precisely at the ecological problematic. With the help of the concept of 're-entry', it permits the formulation of a concept of rationality.[17] Accordingly, a system attains rationality to the extent that it reintroduces the difference of system and environment within the system and is not guided by its (own) identity but by difference. Measured by this criterion, ecological rationality would be attained when society could charge the reactions to its environmental effects to itself. This principle would then have to be reformulated with a corresponding system-reference for every function system in society. It would also have to be noted that all these rationalities could not be added together because every function system calculates its own rationality and treats the rest of society as environment.

There are many reasons for handling this idea carefully – not the least of which is that it is only a scientific theory, i.e., according to its own self-description it is derived from one of the function systems. Only if one realizes this can it also be interpreted as a scientifically considered proposal for a self-observation and self-description of society.[18] No socially thera-peutic effect to counter anxiety can be expected from it. Indeed, it would be highly questionable if relief from pressing problems were to be sought in a theory's narcotic of abstraction, but one should not fear that a sensible handling of systems-theoretical analyses will produce this. They lead more to the expansion of the perspectives of problems than to their suppression. Whoever doubts this ought to go back and read this book a second time. That one should not expect recipes here – as if a net rational improvement could be attained from the closing of nuclear plants or from constitutional reforms effecting a change in the majority rules – is, therefore, self-evident. On the other hand, the alternatives that the rhetoric of anxiety offers enjoy the advantage of being near at hand, even if they are unrealistic. In a way that is almost impossible to explain they obfuscate social interdependencies and the mediation of effects.

One has to concede, however, that both are highly topical possibilities of our society's observation of itself, and one should also like to hope that they could be brought into a communicative relation.

20

Toward a Rationality of Ecological Communication

Anyone who hoped that these reflections on the theme of 'ecological communication' would clarify how this kind of communication could contribute to the solution of pressing environmental problems will be disappointed. Our aim was to work out how society reacts to environmental problems, not how it ought to or has to react if it wants to improve its relation with the environment. Prescriptions of this sort are not hard to supply. All that is necessary is to consume fewer resources, burn off less waste gas in the air, produce fewer children. But whoever puts the problem this way does not reckon with society, or else interprets society like an actor who needs instruction and exhortation (and this error is concealed by the fact that he or she does not speak of society but of persons).

We have also forced a second reservation on ourselves – one regarding criticism. Ordinarily, criticism presupposes that one knows how things are to be improved and then finds fault with the fact that this is not the case. In the reflexive understanding of the Frankfurt School this advance knowledge is no longer assumed but replaced by the dream of the subject or, since Habermas, by the idea that society can determine its own identity in communicative discourse in such a way that this commits the participating subjects 'internally' and binds them to the collective identity. In the self-evidence of the grounds of the validity-claims the collective identity is given in a way that convinces every participant in the communication; and obviously not only in the particularity of the respective actions, situations and grounds but with reference to identity too. If identity were to be interpreted merely as the system's description of itself, the agreement of subjective and collective identity that creates the certainty of the identity's rationality would be lacking. The systems-theory that maintains that 'modern societies cannot develop a rational identity at all lacks any reference point for a critique of modernity'.[1] But this means only that the completed project of modernity, if it is presented in this way, is not carried forward by systems theory. This should not be a bone of contention. The critique of this semantics of modernity says that if this is modern society's ambition then it cannot be grasped sufficiently or, if so, it can only be conceived as failure. It follows that it is hardly imaginable that contemporary society has got into ecological difficulties because it has not taken the project of modernity seriously enough or has not discussed validity-claims and grounds sufficiently.

At the same time, Habermas's suggestion remains true: a centerless society cannot assert a rationality of its own but has to rely on the subsystem rationalities of its function systems. This is also true if the protest directed against it is considered, for this too can only be a partial phenomenon and can neither be nor represent the whole within the whole. At best, it produces a corrosive mistrust without a rationality of its own – to which others, then have to react. From the point of view of systems theory this would have to be the case if rational individuals anywhere came to an agreement about their validity-claims. For why should it be left to them to determine what is good when the starting-point has to be that others do not have valid grounds to disagree with the procedure and even less with the proposed consensus.

The answer to the question of the system's rationality (and also to the question of the rationality of ecological communication) must therefore be a change in the formulation of the problem.[2] In every assumption of a differentiated unity there is inevitably a paradox because the unity of the whole is not outside or above the parts but is identical and not identical with the sum of them at the same time. Besides, we know that it is not possible (or only for specific purposes) to resolve this paradox through a differentiation of levels or a hierarchy of types.[3] One can also show that in modern society every claim of a part to be the whole or to represent identity is subject to observation and contradiction since, for reasons of social structure, there are no longer any non-competitive positions, for example, the apex of the hierarchy or a center *vis-à-vis* a periphery. It is ill-advised and leads to a peculiar Utopianism and hopelessness if one does not recognize these restricting conditions on any attempt at rationality but persists in a belief in direct access to it.

So falls the idea that the environment has a partner in society or even that humanity itself is this partner. This is just a new version of the privilege of representing the whole within the whole because 'the' environment is the correlate of 'the' system and can be seen as a unity only from the unity of the system. Even enthusiasm and the awareness of responsibility cannot privilege anyone in this way. Only as a unity – and this means as a differentiated unity – can society react to its environment. Besides, since none of its individual function systems is organized and capable of making decisions as a unity, an organizational co-ordination is not even attainable.

In addition to these difficulties we are faced with a further problem that the preceding chapters indicated and clarified with the distinction of coding and programming. The unity of every function system resides in being guided by a binary code valid for itself alone. Its unity is its difference, indeed a difference that robs the system of the possibility of placing itself on the 'right' side. This also eliminates teleological rationalities (for example, rationalities of action) that would enable the system to characterize itself as a striving for truth, right, power, wealth, education or the moral life and to be taken, at least as far as good intentions are concerned, as rational. Instead, the difficult questions concerning what really constitutes the unity of the difference guiding such codes and what really constitutes the rationality of a distinction have to be answered. Whatever appears as 'right' within such a code depends on the coded acquisition and processing of information and has meaning in reference to the contingency that is disclosed and structured by this. Only whatever is capable of being its opposite according to a given schema – under the condition of the exclusion of a third alternative – can be right, and since we must assume this schema as functionally specific, a direct inference from it to social rationality is impossible.

If this has been an accurate description of the situation then the problems of the rationality of society as a whole have to be approached in an entirely new manner. Whoever still localizes rationality in the reflexivity of reason – for example, like Habermas, in the reflexivity of a discursively ascertained rationality[4] – will find it impossible from now on to discover rationality either here or in what follows. The traditional conceptual arrangement should not immediately exclude the consideration of how a rational society would have to be conceived if (1) the concept of self-reference (reflexivity) is transferred to all empirical, autopoietic systems; (2) the inference from self-reference to rationality is surrendered; (3) rationality can no longer be found in the self-reference of reason; and (4) all considerations of rationality are required to agree with the paradox of differentiation and the binary coding of the function systems.[5]

Obviously, social rationality can reside neither in the projection of rationality of individual function systems (even science's) nor in its total rejection as irrational. It has to be conceived, as it were, free of any specific location – as a difference that can be realized differently. A sufficiently general, at least hitherto unsurpassed concept for this resides in the generalization of the method of functional analysis when this is conceived as a method for the creation of difference. Its status as rigorously scientific is disputed and is certainly in need of restricting conditions (perhaps from specific theories like systems theory) or from formal requirements (perhaps those of mathematics). The general rule of beginning from reference problems and looking for their functional equivalents can be seen, to a greater extent, as a generalizable principle that accepts unity only as a problem, i.e., only for the sake of the difference that can be created through it. Even with all the practical difficulties involved in finding functional equivalents for everything given, many possibilities of practicing this form of orientation in other function systems, even without the aegis of science, i.e., originarily, become clear.

This changes the focus of all previous theories of reflection from unity to difference and enables them to acquire information in ways that differ from those accepted previously. Until now, the reflection used by function systems, even where it existed in a theoretical form, was guided by the values of correctness and sought the unity of the system in these. Because these theories of reflection are exposed to a functionally oriented comparison and family likenesses that are reducible to the functional differentiation of the social system, a new situation is created that upsets their normative and evaluative self-certainty. A functional reanalysis would have to begin from the system's description of itself within the system. Then it would very quickly realize that every self-description simplifies the system that it describes in terms of a model, i.e., reduces complexity and creates difference: the difference between the system it describes and its description of itself. Every reflection creates an observation of the reflection that works selectively – perhaps in the sense of a recursive sorting-out of the observation's semantic commitments to recall or in the sense of the self-referential systems' 'own values'.[6] These are only examples of theories of reflection about function systems, formed under guide-lines that try to attain a large measure of environmental openness in the function systems and, in this way, represent society for each of them.

Social rationality would naturally require that the ecological difference of society and its external environment is reintroduced within society and used as the main difference. We have to begin from the idea that there can be no privileged place for this, no authoritative organization, consequently no 'constitution' that could transform the ecological difference into binding guide-lines for further information processing. If such a place were to be created the result would be a new internal difference in society: the difference of this place from all the others in society. One can promote this idea as a self-justifying Utopia. But then wouldn't this be just another version of a self-justifying reason? Because of the paradox of a differentiated unity, of a *unitas multiplex*, its implementation would fail for all social formations, and not simply because of functional differentiation. The tautology of rationality – correct is what corresponds to the concept of what is correct – would suddenly turn into a paradox: correct is what is impossible because it presupposes society only as a unity and not as difference. To formulate it more mundanely, this Utopia entails foreseeable difficulties of implementation. This problem is repeated, of course, on all levels of internal social system-formation, but within the subsystems the possibility exists of a hierarchical organization through which the difference of system and environment can be transformed into internal system directives.

These considerations leave the concept of system rationality intact. This signifies the possibility of reintroducing the difference of system and environment within the system, thus the possibility of directing the system's information processing by means of the unity of the difference of system and environment. The unity of the difference of system and environment is the world. Within differentiated systems, however, this reference to the world is filtered not only by the external boundaries of the encompassing system but also by further internal boundaries. This is what the conditions of 'Western rationalization' in Weber's sense hang on: the rationalization of firms and administrations. At the same time these conditions for construction mean that system rationality increasingly loses its claim to be world rationality. Guidance from internal system environments, for example, markets or public opinion, begins to dominate. To the extent that system rationality appears more realizable it becomes less world-rational and even less socially rational, but once this becomes clear one can also see that this is not a matter of an 'iron law' but rather of the costs of increasingly improbable complexity. The precondition of all concern with rationality is the proper understanding why it is – and remains – improbable. Then it makes sense to be guided by the Utopia of rationality: to see whether and how individual systems can be used to provide solutions to problems that are more rational and include further environments. Today it is already clear that communication about ecological themes is beginning to examine such possibilities.

However one assesses these possibilities, it should remain clear that the concept of rationality offered here never designates system-states, therefore it cannot solidify into desired end-states, goals or the like. Nor is it a substantial or a teleological rationality. System rationality is never concerned with unity but with difference and with the resolution of all unity into difference. So we have to ask where the rationality of a distinction arises, considering what can be designated with its help? We suggest that it is out of a reference to the ultimate difference of system and environment, i.e., out of the ecological difference.

21

Environmental Ethics

In conclusion I would like to make some remarks concerning perhaps the most prevalent expectation today – that ecological communication should culminate in ethical questions and find its justification there. In view of the given social situation a change of consciousness is necessary, a new ethics, a new *environmental* ethics. We have already touched on this requirement at different places – and not been able to get very far. Our investigations led us in an entirely different direction. The problem of an environmental ethics cannot be covered sufficiently in a few marginal remarks, so in place of a summary we would like to clarify the difference between a systems-theoretical, sociological analysis and ethics – in the hope that further communication will be guided by this difference and not simply by ethical postulates and maxims.

First, it is important to distinguish morality and ethics (knowing quite well that it is a widespread linguistic custom to use these terms interchangeably). Morality is to be understood as the coding of communication by the binary scheme of good and bad (or, if subjectivized, of good and evil). The code is always applicable when the behavior that is the subject of communication is sanctioned by the bestowal or withdrawal of esteem or contempt.[1] Therefore it can also be said that morality resides in conditioning the attribution of esteem or contempt. In this way 'morality' is an artificial aggregate because it is never necessary or possible for communication that the totality of the conditioning viewpoints be available in a way that sharply distinguishes one from another. The formulation 'morality', therefore, always refers to communication that is already moralized (and, of course, also to non-moralized communication).

It is unavoidable that the observation of morality encounters paradox in its object domain. This is not a specifically moral problem. Instead, as is well known from logic, all binary coding leads to paradox when the code is applied to itself. Sociological analysis discloses this empirically and confirms the practical relevance of paradox in this way.

On one hand, morality seems to have a polemical origin, i.e., it arises out of uncertainty, disunity and conflict because this is the only way of providing an occasion for using esteem or contempt as a sanction. One could even say that morality is concerned with pathological instead of normal cases – including those of supererogatory performances and merits that appear as if this could be expected of everyone – and, therefore, has to be isolated by means of a special esteem given to a hero, martyr, ascete, or virtuoso of virtue.

On the other hand, morality also leads to conflict, whether this is concealed or open. A person committed to morality can relent only with difficulty because his or her self-respect is thereby placed in jeopardy. A person who considers him- or herself to be morally responsible falls very easily into a bind. Such a person must proceed very carefully – except in very limited circumstances when morality (despite its pathological origin) has become self-evident. In the case of dissent this problematic leads very quickly to conflict, notwithstanding the similarly conflictual expedient: that morality requires not only loving the good but also hating and combating the bad.[2]

We have to realize, logically as well as empirically, that morality is paradoxical or, looked at temporally, that it causes paradox. As the unity of the difference of good and bad it works both well and poorly. What is good can be bad and what is bad can be good. Thus the observer of morality finds observation blocked. In any event, the observer cannot make any moral judgement about morality. 'In other words, everything is moral. But morality itself is not moral!'[3]

Only ethics can make a moral judgement about morality. At least it trusts itself to be able to do so. Ethics (no longer in an ethos-related use, thus in a sense that is at present valid) is to be understood as a *reflection theory of morality*. Its function is to reflect the unity of the moral code, the unity of the difference of good and bad. If moral difference raises the problem of its unity (and is not taken simply as nature) it generates ethics. Ethics, therefore, must undertake the task of eliminating the paradox of the moral paradox if it wants to be a moral theory of morality. It can do this only if it does not know what it does because the elimination of the paradox is, of course, a paradoxical undertaking itself. Therefore ethics has to establish an ersatz-problem that enables it to conceal what is primarily of concern. It could not tolerate proceeding, as it were, in a 'carefully careless' way as social morality has prescribed since the seventeenth century. As a theory of reflection it is too obligated to the disclosure of the principle of the unity of difference, so it chooses the unity of a rule that separates the entire domain of morality (good) from itself as the ersatz problem – as in the case of the categorical imperative.

We cannot and do not want to pursue the possibilities of the construction of ethical theories here any longer.[4] We must be content with uncovering the problem in whose avoidance ethics has its latent unity. The secret of ethics, its arcanum, the source of all the experiences that it cannot name and to which it cannot return is the paradox of moral coding. Thus, from the point of view of its function, ethics fails in its task to warn against morality. This is left to a much overworked sociology. But here, too, an observing of observing, a second-order observing is required.

These considerations are generally valid – both for earlier experiences with a social ethics that derives its problem from humanity's treatment of humanity as well as for a possible environmental ethics that resorts to moral conditioning, i.e., to moral resonance. We will also have to consider whether the problems of paradox and its avoidance do not become more acute when viewed from the perspective of environmental ethics, but since this ethics does not exist at present as a theory of reflection this will be hard to tell. We cannot even exclude the possibility that all ethical reflection fails because of the non-moral nature of particular problems of risk. But we will not consider this. Even so, we still have to realize that the discrepancy between social problems and environmental problems will make itself known – formulated according to systems theory, this means discrepancies between the society and its environment. In the case of morality and ethics the concern, naturally, is with a *social* regulation, but precisely because of this we will have to ask whether the conditions and forms of this regulation do not have to change if they are extended to an unrelated domain, to non-social sources of problems. It would be premature, however, to take this question out of circulation by saying that even ecological problems are ultimately caused socially or, at least, are of interest only in the context of ecological communication. However correct this may be, an entirely new dimension of complexity comes into play through the difference of system and environment, and it is improbable that this complexity – just like the internal social complexity of double contingency – could be transferred to the conditions of esteem or contempt.

We must, therefore, ask the question whether, even under these circumstances, an ethics that abstains from paradox will develop and be able to be practiced with moral responsibility. This could also give us pause to wonder whether it is not the recognition of paradox that is the way for ethics to go to do justice to the new problem situation – for, even in the case of theories, a more complex problem situation changes the conditions of adequate internal complexity. It could very well be that the digestive as well as the ruminating apparatus of ethics will have to be equipped with more stomachs – above all with one for paradoxes.

In any event, as long as such an ethics does not exist, ecological communication itself will have to respect its distance from morality. At present it is falsely guided by the instructions of environmental ethics. To be sure, ecological communication will examine ethical possibilities too and perhaps be able to prepare a field of development for their reformulation. But if anywhere, it is in ecological communication that society places itself in question, and we cannot see how ethics can dispense with this and remain available as something that can be relied on. On the contrary, if a specific function is to be attributed to environmental ethics within the context of ecological communication then this might very well be to remain cautious in dealing with morality.

Glossary

The text uses a series of concepts in a way that is peculiar to it and with a precision that depends on complex preliminary considerations. Since a sufficient justification for the use of these concepts was not possible within the text I will present some definitions with brief explanations here.

Autopoiesis Refers to (autopoietic) systems that reproduce all the elementary components out of which they arise by means of a network of these elements themselves and in this way distinguish themselves from an environment – whether this takes the form of life, consciousness or (in the case of social systems) communication. Autopoiesis is the mode of reproduction of these systems.

Code Codes arise out of a positive and a negative value and enable the transformation of the one into the other. They come into being through a duplication of a given reality and with this offer a scheme for observations within which everything that is observed appears as contingent, i.e., as possibly different.

Communication Designates not simply an act of utterance that 'transfers' information but an independent autopoietic operation that combines three different selections – information, utterance and understanding – into an emergent unity that can serve as the basis for further communication.

Complexity A state of affairs is complex when it arises out of so many elements that these can only be related to one another selectively. Therefore complexity always presupposes, both operatively as well as in observation, a reduction procedure that establishes a model of selecting relations and provisionally excludes, as mere possibilities, (i.e., potentializes) other possibilities of connecting elements together.

Coupling This concept designates the reciprocal dependency of system and environment which can be seen by an observer if the latter takes the distinction of system and environment as basic. The observer can even be the system itself if it is in the position to observe itself when it uses the distinction of system and environment.

Differentiation, functional In the text this concept refers to the formation of systems within systems. It does not necessarily designate the decomposition of an entire system into subsystems but rather the establishment of system/environment differences within systems. The differentiation is functional in so far as the subsystem acquires its identity through the fulfillment of a function for the entire system.

Ecology Means in this context the totality of scientific investigations that concern themselves, on whatever level of system formation, with the consequences of the differentiation of system and environment for the system's environment. The concept does not presuppose any specific kind of system (ecosystem).

Elimination, of the paradox see **Paradox**.

Observation Is defined on the level of abstraction of the concept of autopoiesis. It designates the unity of an operation that makes a distinction in order to indicate one or the other side of this distinction. Its mode of operation can, again, be life, consciousness or communication.

Paradox A paradox occurs when the conditions of the possibility of an operation are at the same time the conditions of the impossibility of this operation. Since all self-referential systems having the possibility of negating create paradoxes that block their own operations (for example, can determine themselves only in reference to what they are not, even if they themselves and nothing else are this non-being) they have to foresee possibilities of *eliminating the paradox* and at the same time disguise the operations necessary for this. For example, they have to be able to treat the recursive symmetry of their self-reference asymmetrically, either temporally or hierarchically, without being able to admit to themselves that an operation of the system itself is necessary for this transformation.

Program Refers to that of codes and, following a well-established conceptual usage (canon, criterion, regula), designates the conditions under which the positive or negative value of a specific code can be ascribed to situations or events. In social systems this is treated as a question of a decision (thus also decision-programs) between true and false, legal and illegal etc.

Redundancy The multiple certification of a function, therefore the appearance of 'superfluity'. The rejection of redundancy means that multifunctional mechanisms have to be replaced by functionally specific ones that are applied to (autopoietic) self-certification.

Representation Is used to designate the presentation of the unity of a system by a part of it (*repraesentatio identitatis*) and is distinct from the rigorously legal sense that has legally effective substitution in mind. In so far, representation is always paradoxical: intending to present unity, it creates a difference between the representing and the other parts of the system.

Resonance Signifies that systems can react to environmental events only in accordance with their own structure.

Self-reference Designates every operation that refers to something beyond itself and through this back to itself. Pure self-reference that does not take this detour through what is external to itself would amount to a tautology. Real operations or systems depend on an 'unfolding' or de-tautologization of this tautology because only then can they grasp that they are possible in a real environment only in a restricted, non-arbitrary way.

Social Systems A social system comes into being whenever an autopoietic connection of communications occurs and distinguishes itself against an environment by restricting the appropriate communications. Accordingly, social systems are not comprised of persons and actions but of communications.

Society   That social system which includes all meaningful communication and is always formed when communication takes place in connection with earlier communication or in reference to subsequent communication (i.e., autopoietically).

Notes

Translator's Introduction

1. Niklas Luhmann, *Soziale Systeme* (Social Systems), Frankfurt: Suhrkamp, 1984.
2. Jürgen Habermas/Niklas Luhmann, *Theorie der Gesellschaft oder Sozialtechnologie: was leistet die systemforschung?* (Theory of Society or Social Technology: What Does Systems Research Accomplish?), Frankfurt: Suhrkamp, 1971.
3. Talcott Parsons, *Social Systems and the Evolution of Action Theory*, New York: Free Press, 1977, p. 118.
4. Talcott Parsons et al., *Toward a General Theory of Action*, Cambridge, Mass.: Harvard University Press, 1951, p. 16.
5. Cf. *Soziale Systeme*, 1984, p. 93.
6. Hessicher Rundfunk, 7 May 1987.

Preface

1. The text of the address has been published in the Proceedings of the Academy under the given title, RWAkW G278, Opladen 1985.

Chapter 1 Sociological Abstinence

1. 'I write out of a sense of alarm,' admits John Passmore, *Man's Reponsibility for Nature: Ecological Problems and Western Traditions*, New York 1974, p. IX, and he undoubtedly speaks for many ecological writers.
2. For a summary cf., Josef Müller, 'Umweltveränderungen durch den Menschen', in Karl Heinz Kreeb, *Ökologie und menschliche Umwelt: Geschichte – Bedeutung – Zukunftsaspekte*, Stuttgart 1979, pp. 8–69.
3. Cf., the last two chapters in Amand L. Mauss, *Social Problems as*

Social Movements, Philadelphia 1975, which are preceded by an extensive enumeration of problem domains.

4. In unpublished writings on the early history of liberalism Stephen Holmes calls this 'antonym substitution'.

5. For the absence of a developed 'milieu' concept in the eighteenth century cf., Georges Canguillem, *La Connaissance de la vie*, 2nd edn, Paris 1965, pp. 129ff. Cf., also Jürgen Feldhoff, 'Milieu', in *Historisches Wörterbuch der Philosophie*, vol. 5, Basel-Stuttgart 1980, pp. 1393–5; and Leo Spitzer, '"Milieu" and "Ambience": An Essay in Historical Semantics' in *Philosophy and Phenomenological Research*, vol. 3 (1942), pp. 1–42, 169–218.

6. Cf., Richard Hofstadter, *Social Darwinism in American Thought 1860–1915*, Philadelphia 1945; Emerich K. Francis, 'Darwins Evolutionstheorie und der Sozialdarwinismus', in *Kölner Zeitschrift für Soziologie und Sozialpsychologie*, vol. 33 (1981), pp. 209–28; Niles Eldrege/Ian Tattersall, *The Myths of Human Evolution*, New York 1982; Walter Bühl, 'Gibt es eine soziale Evolution?', in *Zeitschrift für Politik*, vol. 31 (1984), pp. 302–32.

7. Programmatically in 'Zu einer allgemeinen Systemlehre', *Biologia Generalis*, vol. 19 (1949). pp. 114–29, and with a far-reaching effect in the context of the English language. Cf., Ludwig von Bertalanffy, *General Systems Theory: Foundation, Development, Applications*, London 1971. For a historical estimation cf., also I. V. Blauberg/V. N. Sadovsky/E. G. Yudin, *Systems Theory: Philosophical and Methodological Problems*, Moscow 1977.

8. Cf., for example, Walter Buckley, *Sociology and Modern Systems Theory*, Englewood Cliffs, N.J. 1967; Kenneth F. Berrien, *General and Social Systems*, New Brunswick N.J. 1968. In the important anthology that documents the trend – Walter Buckley (ed.), *Modern Systems Research for the Behavioral Scientist*, Chicago 1968 – sociologists are hardly mentioned. Anyone familiar with the theoretical history of sociology knows that this is connected with the dominating influence of the structurally functional paradigm of that time. Its main representative, Talcott Parsons, had accepted some suggestions from general systems theory but developed a theory in which environments played a role only as systemically 'internal'.

9. Especially since Tom Burns/G. M. Stalker, *The Management of Innovation*, London 1961; and Paul R. Lawrence/ Jay W. Lorsch, *Organization and Environment: Managing Differentiation and Innovation*, Boston 1967. Among the many textbooks cf., also

Howard E. Aldrich, *Organizations and Environments*, Englewood Cliffs, N.J. 1979.

10. Moreover, the same is true for the theory of the political system, above all for David Easton, *A Systems Analysis of Political Life*, New York 1965.

11. Walter L. Bühl, 'Das ökologische Paradigma in der Soziologie', in Harald Niemeyer (ed.), *Soziale Beziehunsgeflechte: Festschrift für Hans Winkmann*, Berlin 1980, pp. 97–122; and Bühl, *Ökologische Knappheit: gesellschaftliche and technologische Bedingungen ihrer Bewaltigung*, Göttingen 1981, p. 35, emphasizes this in view of the superficiality of previous systems analyses. Of course, with very different possibilities aimed at the ecosystem in mind.

12. Cf., Niklas Luhmann, *Soziale Systeme: Grundriss einer allgemeinen Theorie*, Frankfurt 1984.

13. Cf., Lynne White Jr, 'The Historical Roots of Our Ecological Crisis', in *Science*, vol. 155 (1967), pp. 1203–7; reprinted with critical replies in Ian G. Barbour (ed.), *Western Man and Environmental Ethics: Attitudes Toward Nature and Technology*, Reading Mass. 1973, pp. 18–30.

14. Cf., Günther Altner, 'Ist die Ausbeutung der Natur im christlichen Denken begründet?' in Hans Dietrich Englehardt et. al. (eds), *Umweltstrategie: Materialien und Analysen zu einer Umweltethik der Industriegesellschaft*, Gütersloh 1975, pp. 33–47; Robin Attfield, 'Christian Attitudes to Nature', in *Journal of the History of Ideas*, vol. 44 (1983), pp. 369–86; Attfield, *The Ethics of Environmental Concern*, Oxford 1983.

15. A typical argument for the dominance of ethical viewpoints, 'results because the exclusivity of scientific, technological and economic viewpoints have led to the crisis (*sic!*) described above' (according to Heinhard Steiger, 'Begriff und Geltunsebenen des Umweltrechts', in Jürgen Salzwedel (ed.), *Grundzüge des Umweltrechts*, Berlin 1982, pp. 1–20, vol. 13 – as if such an exclusivity had always existed! But even the sophisticated analyses of Hans Jonas, *Das Prinzip Verantwortung: Versuch einer Ethik für die technologischen Zivilisation*, Frankfurt 1979, has inadequate historical analyses because of its desire to make a clear contrast.

16. This remarkable inconsistency has also occurred to Louis Dumont. Cf., *Essai sur l'individualisme: Une perspective anthropologique sur l'idéologie moderne*, Paris 1983, p. 203.

17. It produces considerable confusion when a widespread linguistic custom, whose consistent application would have to lead to

economizing on the use of the concept of ecology, designates ecological interdependencies or 'balances' as 'system' (ecosystem). For example, according to Heinz Ellenberg, 'Ziele und Stand der Ökosystemforschung', in Ellenberg (ed.), *Ökosystemforschung*, Berlin 1973, pp. 1–31, vol. 1, 'An eco-system is a context of interaction [*Wirkungsgefüge*] including living beings and their inanimate environment that is open but also capable of self-regulation to a certain degree . . . Ecological systems are always open, i.e., capable of being disturbed from outside and without sharp boundaries.' The same is true according to Kreeb, ibid., (1979) and the prevailing opinion. But not every interconnection is a system. A system exists only when an interconnection distinguishes itself from an environment. (Exactly the opposite of Bühl, ibid., 1980, p. 121. In this sense, for example, one can speak of the physical system of the planet Earth in which the human organism, the vocal transmission of human communication, the microphysics of the human ear, etc. participate. This designates a systems-theoretical but not an ecological problematic. In distinction to a simple systems-theoretical problematic, a problematic is called ecological only when it aims at unity despite difference or even at unity through difference, i.e., because a system/environment interconnection is structured through the system separating itself from its environment, differentiating itself from it and on this basis developing a highly selective behavior toward it. So the ecological problematic cuts across the systems-theoretical problematic (which, of course, does not exclude that investigations in the one perspective cannot be relevant for the other). For the ecology of human society countless systems are relevant (perhaps the closed system of genetic inheritance) without the unity of this system and its environment identifying with the ecology of society, i.e., with the system/environment relation of society. At present it is an open question, discussed under the title of 'sociobiology', whether and to what extent the handing-down of traits is relevant for the system of society and its environment.

18. Cf., Spitzer, ibid., especially the first part.
19. Theoretical motifs like this that require a boundary line in the world for the self-observation and reflection of the world have appeared in the philosophy of reflection and in cybernetics in connection with Wittgenstein. Cf., for example, Gotthard Günther, 'Cybernetic Ontology and Transjunctional Operations', in Günther, *Beiträge zur Grundlegung einer operationsfähigen Dialektik*, vol. 1, Hamburg 1976, pp. 249–328 (especially pp. 318ff.). Sociology

has only reached the point of self-fulfilling or self-defeating prophecies.

Chapter 2   Causes and Responsibilities?

1. In a somewhat different sense – in reference to problems of scarcity and their temporal operationalizations – Guido Calabresi/Philip Bobbitt, *Tragical Choices*, New York 1978, speak of tragic decisions. I think it comes closer to the classical concept of what is tragic if one turns attention to the participation in causality. One can always find the ultimate 'ground' of the tragic in its providing 'too few' causal possibilities.

2. Cf., for example, Eckhard Rehbinder, *Politische und rechtliche Probleme des Verursacherprinzips*, Berlin 1973; Dieter Cansier, 'Die Förderung des umweltfreundlichen technischen Fortschritts durch die Anwendung des Verursacherprinzips', in *Jahrbuch für Sozialwissenschaft*, vol. 29 (1978), pp. 145–63; Robert Weimar, 'Zur Funktionalität der Umweltgesetzgebung im industriellen Wachstumsprozess', in *Festschrift Bruno Gleitze*, Berlin 1978, pp. 511–26 (519ff.). To a great extent jurists focus on prejudices about attribution and the real problematic arises for them only with the question of whether there should be a further restriction to additional legal consequences, perhaps by raising the issue of guilt. This, of course, is unavoidable if punishment comes into consideration. For economists, however, it is clear that the causer-principle is easy to regulate technically but does not work optimally for allocation. In a certain way, this reservation, too, indicates that attribution rests on a simplification.

3. The selection is guided by 'how the greatest environmental quality can be attained and which procedure appears as the economically and administratively favorable solution,' according to Eckard Rehbinder, 'Allgemeines Umweltrecht', in Jürgen Salzwedel (ed.), *Grundzüge des Umweltrechts*, Berlin 1982, pp. 81–115 (96ff.). Cf., also Rehbinder, ibid., (1973), pp. 33ff. In other words, the causer is whoever one can catch.

4. This means, for example, that the discussion of the idea that 'capitalism' and the unrestricted use of the profit motive are the real causes of environmental damage is just as correct or incorrect as any one-factor theory. Cf., for example, Gerhard Kade, 'Umwelt: Durch das Profitmotiv in die Katastrophe', in Regina Molitor (ed.). *Kontaktstudium Ökonomie und Gesellschaft*, Frankfurt 1972, pp.

237–47, or the contributions of Gerhard Kade and Volker Ronge in Manfred Glagow (ed.), *Umweltgefährdung und Gesellschaftssystem*, Munich 1972. It needs no particular mention that greatly differentiated points of departure for ecological analyses are to be found in the work of Karl Marx. Cf., for example, Peter A. Victor, 'Economics and the Challenge of Environmental Issues', in Herman Daly (ed.), *Economy, Ecology, Ethics: Essays Towards a Steady-State Economy*, San Francisco 1980, pp. 194–240 (207ff.).

5. This is very clear in Heinz von Foerster, 'Cybernetics of Cybernetics', in Klaus Krippendorff (ed.), *Communication and Control in Society*, New York 1979, pp. 5–8.

6. Cf., Walter Benjamin, 'Zur Kritik der Gewalt', in Benjamin, *Gesammelte Schriften*, vol, II.1, Frankfurt 1977, pp. 179–203.

7. Cf., for example, the reduction of ecological problems to the relation of scarcity and allocation, in Horst Siebert, *Ökonomische Theorie der Umwelt*, Tübingen 1978.

Chapter 3   Complexity and Evolution

1. Quoted from Stafford Beer, *Designing Freedom*, New York 1974, pp. 7, 10, 95.

2. More exactly, this means that there is no 'requisite variety' for *this* difference; or in other words, no system can acquire enough complexity of its own to be able to control the complexity in its environment. Of course, this does not exclude the possibility of planning models, machines or systems that have requisite variety for the states of affairs to be controlled.

3. Cf., Niklas Luhmann, *Soziale Systeme*, Frankfurt 1984, pp. 47ff. and 249ff.

4. This is a theoretical approach that has been used frequently in recent times. Cf., for example, Gerhard E. Lenski, 'Social Structure in Evolutionary Perspective', in Peter M. Blau (ed.), *Approaches to the Study of Social Structure*, London 1976, pp. 135–53; Philippe Van Parijs, *Evolutionary Explanation in the Social Sciences: An Emerging Paradigm*, London 1981; Bernhard Giesen/ Christoph Lau, 'Zur Anwendung Dawinistischer Erklärungsstrategien in der Soziologie', in *Kölner Zeitschrift für Soziologie und Sozialpsychologie*, vol. 33 (1981), pp. 229–56; Michael Schmid, *Theorie sozialen Wandels*, Opladen 1982.

5. Among others cf., André Béjin, 'Différenciation, complexification,

évolution des sociétés', in *Communications*, vol. 22 (1974), pp. 105–18.

6. This is already incontestable for the concept of complexity. Cf., for example, Todd R. La Porte, 'Organized Social Complexity: Explication of a Concept', in La Porte (ed.), *Organized Social Complexity: Challenge to Politics and Policy*, Princeton, N.J. 1975, pp. 3–39.

7. An increasing criticism of this idea can be observed over the last few years through the development of theories of the self-organization of thermodynamically open systems and the self-referential formation of systems. Cf., for example, Edgar Morin, *La Méthode*, vol. 2, Paris 1980, pp. 47ff.; Alfred Gierer, 'Socioeconomic Inequalities: Effects of Self-Enhancement, Depletion and Redistribution', in *Jahrbuch für Nationalökonomie und Statistik*, vol. 196 (1981), pp. 309–31; Gerhard Roth, 'Conditions of Evolution and Adaption in Organisms as Autopoietic Systems', in D. Mossakowski/ G. Roth (eds), *Environmental Adaption and Evolution*, Stuttgart 1982, pp. 37–48.

8. Today, a sceptical, or at least a very circumspect opinion predominates in this respect. For social systems cf., for example, Mark Granovetter, 'The Idea of "Advancement" in Theories of Social Evolution and Development', in *American Journal of Sociology*, vol. 85 (1979), pp. 489–515; Walter L. Bühl, 'Gibt es eine soziale Evolution?' in *Zeitschrift für Politik*, vol. 31 (1984), pp. 302–32.

9. A standard interpretation that advises conserving the developmental potential of whatever is unspecific. Cf., for example, E. D. Cope, *The Primary Factors of Organic Evolution*, Chicago 1896, pp. 172ff.; Elman R. Service, *The Law of Evolution and Culture*, Ann Arbor, Mich. 1960, pp. 93ff.; also in Elman R. Service, *Cultural Evolutionism: Theory in Practice*, New York 1971, pp. 31ff.

Chapter 4   Resonance

1. Cf., Humberto Maturana, *Erkennen: Die Organisation und Verkörperung von Wirklichkeit*, Braunschweig 1982, for example, pp. 20ff., 150ff., 287ff.; Francisco Varela, 'L'auto-organisation: de l'apparence au mécanisme', in Paul Dumouchel/Jean-Pierre Dupuy (eds), *L'auto-organisation: de la physique au politique*, Paris 1983, pp. 147–64 (148).

2. For a more detailed analysis cf., the chapter on meaning in Niklas

Luhmann, *Soziale Systeme*, pp. 92ff.

3. 'Indeterminacy means the necessary determinacy of a strictly prescribed style,' as is said in *Ideen zu einer reinen Phänomenologie und phänomenologischen Philosophie*, vol. 1, Husserliana vol. III, The Hague 1950, p. 100.

4. Cf., Francisco Varela, *Principles of Biological Autonomy*, New York 1979; Maturana, ibid. (1982).

5. The theoretical sources of this idea are quite heterogeneous and difficult to survey. The neo-dialectical tradition, especially Hegel, comes readily to mind. Cf., also Ferdinand de Saussure, *Cours de linguistique générale*, 5th edn, Paris 1962; Alfred Korzybski, *Science and Sanity: An Introduction to Non-Aristotelian Systems and General Semantics*, 4th edn, Lakeville Conn. 1958; George A. Kelly, *The Psychology of Personal Constructs*, 2 vols, New York 1955.

6. Cf., above chapter 1, note 19.

7. Cf., for this Helmut Willke, 'Zum Problem der Intervention in selbstreferentielle Systemen', in *Zeitschrift für systemische Therapie*, vol. 2 (1984), pp. 191–200.

8. Cf., for example, Korzybski, ibid., pp. 386ff.

9. For the semantics of totalitarianism that tries to negate this cf., Marcel Gauchet, 'L'Expérience totalitaire et la pensée de la politique', in *Esprit*, July/August 1976, pp. 3–28.

Chapter 5   The Observation of Observation

1. Cf., Roy A. Rappaport, *Ecology, Meaning and Religion*, Richmond Ca. 1979, pp. 97ff.

2. Cf., Maturana, ibid. (1982), especially pp. 36ff. Maturana calls the other-reference of first-order observation 'niches' and reserves 'environment' for that which reveals itself to second-order observation as the other-reference of the observed system. We have not followed this terminology in the text because, although it is univocal and clear, it would force us to use a terminology that constantly deviates from the standard usage.

3. The cybernetic theory of Heinz von Foerster is based on this viewpoint. Cf., *Observing Systems*, Seaside Ca. 1981.

4. This is the well-known argument of Douglas R. Hofstadter, *Gödel, Escher, Bach: An Eternal Golden Braid*, Hassocks, Sussex 1979.

5. Thus in the form of a question for which there is still no answer, Lars Löfgren, 'Some Foundational Views on General Systems and

the Hempel Paradox', in *International Journal of General Systems*, vol. 4 (1978), pp. 243–53 (244).

6. Cf., in particular 'On Constructing a Reality', in Heinz von Foerster, ibid., (1981), pp. 288–309, and 'Objects: Tokens for (Eigen-)Behaviors', in ibid., pp. 274–85. Cf., also John Richards/ Ernst von Glasersfeld, 'Die Kontrolle von Wahrnehmung und die Konstruktion von Realität', in *Delfin III* (1984), pp. 3–25.

7. Cf., again Heinz von Foerster, 'Cybernetics of Cybernetics', in ibid.

8. See for this Edward E. Jones/Richard E. Nisbett, 'The Actor and the Observer: Divergent Perceptions of the Causes of Behavior', in Edward E. Jones et al., *Attribution: Perceiving the Causes of Behavior*, Morristown N.J. (1971), pp. 79–94, and as a new survey of this 'fundamental attribution error' – i.e., the neglect of situational factors – Lee Ross/Craig A. Anderson, 'Shortcomings in the Attribution Process', in David Kahnemann/Paul Slovic/Amos Tversky (eds), *Judgement under Uncertainty: Heuristics and Biases*, Cambridge, UK 1982, pp. 129–52 (135ff.), or in a more developed way, Francesco Pardi, *L'osservabilità dell'agire sociale*, Milan 1985. Social psychology noticed immediately that it also characterized itself through this insight: as the observer who must also realize that the observed actor follows different principles of attribution than his observer. Cf., for this Wulf-Uwe Meyer/Heinz-Dieter Schmalt, 'Die Attributionstheorie', in D. Frey (ed.), *Kognitive Theorien der Sozialpsychologie*, Bern 1978, pp. 98–136. See further, the interesting essay of Walter Mischel, 'Toward a Cognitive Social Learning Reconceptualization of Personality', in *Psychological Review*, vol. 80 (1973), pp. 252–83.

9. Philippe Van Parijs, *Evolutionary Explanation in the Social Sciences: An Emerging Paradigm*, London 1981, pp. 129ff., calls this the 'principle of suspicion' and notes that this kind of analysis encounters an 'authoritative self-knowledge' that it cannot resolve into suspicion. This, too, is a variant of the general difference of first- and second-hand observation.

10. If one reads the recent work of Jürgen Habermas, especially *Der philosophische Diskurs der Moderne*, Frankfurt 1985, guided by these considerations then it appears as a critique of the critique of the self-descriptions of modern society, i.e., as a kind of third-order cybernetics in the environment specific for it: literature. It is consistent then to carry out the discourse as the discussion of opinions that authors have expressed about other authors (Hegel about Kant, Heidegger about Nietzsche, etc.). The pristine

transparency of these presentations can be gained only on the basis of an extreme reduction of one's own theory that dismisses the aporias of a self-clarifying reason and merely requires that one presents testable validity claims in communication. Through this simplification the description of the description of descriptions acquires a considerable succinctness but at the same time an unbridgeable distance from real social operations that are then indirectly transfigured as life-world.

Chapter 6 Communication as a Social Operation

1. Thus, for example, with unproductive explanatory remarks, Eric Trist, 'Environment and Systems-Response Capability', in *Futures*, vol. 12 (1980), pp. 113–27.
2. Ibid., 1983.
3. Cf., Niklas Luhmann, 'Autopoiesis des Bewusstseins', in *Soziale Welt*, forthcoming.
4. This has consequences for the 'de-subjectivization' of the concept of communication that I have worked out elsewhere. See *Soziale Systeme*, ibid., pp. 191ff.

Chapter 7 Ecological Knowledge and Social Communication

1. Cf., for an older survey of the literature June Helm 'Ecological Approach in Anthropology', in *American Journal of Sociology*, vol. 67 (1962), pp. 630–9. Cf., also Julian H. Steward, *Evolution and Ecology, Essays on Social Transformation*, Urbana Ill. 1977; Roy A. Rappaport, *Ecology, Meaning and Religion*, Richmond Ca. 1979. The question of ecological self-regulation that is important for our comparison is distinguished from the predominant problematic that asks whether and to what extent ecological conditions can explain differential evolution, i.e., evolutionary progress as well as the retardation of social development. Today, this kind of theory finds itself exposed to many critical objections; cf., for example, Elman R. Service, *Primitive Social Organization: An Evolutionary Perspective*, New York 1962, pp. 65ff., 72ff.; Robert L. Winzeler, 'Ecology, Culture, Social Organization, and State Formation in Southeast Asia', in *Cultural Anthropology*, vol. 17 (1976), pp. 623–32. Kent V. Flannery, 'The Cultural Evolution of Civilizations', in *Annual Review of Ecology and Systematics*, vol. 3 (1972), pp. 399–426, pleads for explanatory models that

are more complex (including the processing of cultural information). This line of discussion also confirms that one has to understand society as an operatively closed, but self-reactive and thereby environmentally open system.

2. Cf., Roy A. Rappaport, *Pigs for the Ancestors*, New Haven 1968.
3. Cf., again for New Guinea, Frederik Barth, *Ritual and Knowledge among the Baktaman of New Guinea*, Oslo 1975.
4. Cf., Walter J. Ong, *The presence of the Word: Some Prolegomena for Cultural and Religous History*, New Haven 1967; Ong, *Rhetoric, Romance and Technology: Studies in the Interaction of Expression and Culture*, Ithaca N.Y. 1971; Ong, *Interfaces of the World: Studies in the Evolution of Consciousness and Culture*, Ithaca N.Y. 1977.
5. *De libero arbitrio*, Ia 7ff., especially 10; quoted according to *Ausgewählte Schriften*, (ed. Werner Welzig), vol 4, Darmstadt 1969, pp. 11ff.
6. Cf. A. J. Festugière, *La révélation d'Hermes Trismégiste*, 4 vols, Paris 1950–4; Frances Yates, *Giordano Bruno and the Hermetic Tradition*, Chicago 1964.
7. Cf. Micahel Giesecke, 'Überlegungen zur sozialen Funktion und zur Struktur handschriftlicher Rezepte im Mittelalter', in *Zeitschrift für Literaturwissenschaft und Linguistik*, vols 51/52 (1983), pp. 167–84.
8. In reference to tribal cultures Rappaport, ibid., (1979), pp. 100ff., says 'Because knowledge can never replace respect as a guiding principle in our ecosystemic relations, it is adaptive for cognized models to engender respect for that which is unknown, unpredictable, and uncontrollable, as well as for them to codify empirical knowledge.'
9. Thus Thomas Wright, *The Passions of the Minde in Generall*, rev. ed, London 1630, reprinted Urbana Ill. 1971, p. 141.
10. Cf., also Niklas Luhmann, *Die Funktion der Religion*, Frankfurt 1977, especially pp. 255ff.; Niklas Luhmann/Karl Eberhard Schorr, *Reflexionsprobleme im Erziehungssystem*, Stuttgart 1979, pp. 24ff.; Luhmann, *Gesellschaftstruktur und Semantik*, vol. 1, Frankfurt 1980, pp. 9ff.; Luhmann, *Politische Theorie in der Wohlfahrtstaat*, Munich 1981, pp. 19ff.; Luhmann, 'Gesellschaftsstrukturelle Bedingungen und Folgeprobleme des naturwissenschaftlich-technischen Fortschritts', in Reinhard Löw et al. (ed.), *Fortschritt ohne Mass?*, Munich 1981, pp. 113–31; Luhmann, *The Differentiation of Society*, New York 1982, pp. 229ff.; Luhmann, 'Anspruchsinflation im Krankheitssystem: Eine Stellungnahme aus gesellschafts-

theoretischer Sicht', in Philipp Herder-Dorneich/Alexander Schuller (eds), *Die Anspruchsspirale*, Stuttgart 1983, pp. 168–75; Luhmann, 'Die Wirtschaft der Gesellschaft als autopoietisches System', in *Zeitschrift für Soziologie*, vol. 13 (1984), pp. 308–27.

Chapter 8   Binary Coding

1. Cf., for example, Achille Adigò. *Crisi di governabilità e mondi vitale*, Bologna 1980; Jürgen Habermas, *Theorie des kommunikativen Handelns*, Frankfurt 1981, vol. 2, pp. 171ff.
2. Cf., for this chapter 8 note 5 below.
3. See especially for the code of truth Niklas Luhmann, 'Die Ausdifferenzierung von Erkenntnisgewinn: Zur Genese von Wissenschaft', in Nico Stehr/Volker Meja (eds), *Wissenschaftssoziologie*, special volume 22/1980 of the *Kölner Zeitschrift für Soziologie und Sozialpsychologie*, Opladen 1981, pp. 102–39.
4. Antecedents to this are the ancient custom of designating totalities with dual expressions – like 'heaven and hell' or 'court and country'. Cf., for example, Ernst Kemmer, *Die polare Ausdrucksweise in der griechischen Literatur*, Würzburg 1903; Adhemar Massart, 'L'Emploi, en égyptien, de deux termes opposés pour exprimer la totalité', in *Mélanges bibliques*, Paris 1957, pp. 38–46; G. E. R. Lloyd, *Polarity and Analogy: Two Types of Argumentation in Early Greek Thought*, Cambridge, UK 1966; Louis Dumont, *Homo Hierarchicus: The Caste System and its Implications*, London 1970, especially pp. 42ff.
5. Michel Serres, *Le Parasite*, Paris 1980.
6. Even for dualisms in earlier social formations it was true that they were presented, typically, to indicate more than only one and used or not depending on the situation. Only in this way was the exclusion of third possibilities and the proscription of mixed forms possible. Cf., among the extensive literature, for example, Edmund Leach, 'Anthropological Aspects of Language: Animal Categories and Verbal Abuse', in Eric E. Lenneberg (ed.), *New Directions in The Study of Language*, Cambridge, Mass. 1964, pp. 23–63; Mary Douglas, *Purity and Danger: An Analysis of the Concepts of Pollution and Taboo*, London 1966, especially pp. 162ff.; Victor Turner, *The Ritual Process: Structure and Anti-Structure*, London 1969, pp. 38ff.; Rodney Needham (ed.), *Right and Left: Essays on Dual Symbolic Classification*, Chicago 1973.
7. Cf., for an analysis of such matters within the legal system Niklas

Luhmann, 'Die Theorie der Ordnung und die natürlichen Rechte', in *Rechtshistorisches Journal*, vol. 3 (1984), pp. 133–49.

8. Plato, *Lysis*, 215 E.

9. Even 300 years ago this meant that, 'To the Royal Society it will be at any time almost as acceptable to be confuted, as to discover,' according to Thomas Sprat, *The History of the Royal Society of London, For the Improving of Natural Knowledge*, London 1667, reprint St Louis-London 1959, p. 100.

10. Especially for the literature on law and jurisprudence. Cf., for example, Josef Esser, *Vorverständnis und Methodenwahl in der Rechtsfindung: Rationalitätsgarantien der richterlichen Entscheidungspraxis*, Frankfurt 1970; Philippe Nonet/Philip Selznick, *Law and Society in Transition*, London 1979; Gunther Teubner, 'Reflexives Recht: Entwicklungsmodelle des Rechts in vergleichender Perspektive', in *Archiv für Rechts- und Sozialphilosophie*, vol. 68 (1982), pp. 13–59.

11. According to Gregory Bateson, *Ökologie des Geistes: Anthropologische, biologische und epistemologische Perspektiven*, Frankfurt 1981, especially pp. 515ff.

12. For the belief that alphabetized writing could have been the stimulus for this cf., Jack Goody/Ian Watt, 'The Consequences of Literacy', in *Comparative Studies in Society and History*, vol. 5 (1963), pp. 305–45. Another explanation could refer to the high degree of the Greek state's structural differentiation and to the already widespread 'privatization' of religious participation. Cf., for this S. C. Humphreys, 'Approaches to the Study of Structural Differentiation', in J. Friedman/M. J. Rowlands (eds), *The Evolution of Social Systems*, Pittsburgh 1978, pp. 341–71.

13. Distinctions like center/periphery and great tradition/little tradition refer to this. Cf., Edward Shils, 'Centre and Periphery', in *The Logic of Personal Knowledge: Essays presented to Michael Polanyi*, London 1961, pp. 117–31; Robert Redfield, *Peasant Society and Culture: An Anthropological Approach to Civilization*, Chicago 1956.

Chapter 9 Codes, Criteria, Programs

1. See, for example, Karl R. Popper, *Objective Knowledge: An Evolutionary Approach*, Oxford 1972, pp. 13, 317ff.

2. See in regard to truth Sextus Empiricus, *Adversos Mathematicos*,

II 80, quoted according to *Opera* vol. III, Leipzig (Teubner), no date, p. 100.

3. See also the corresponding distinction of values and programs in Niklas Luhmann, *Rechtssoziologie*, 2nd edn, Opladen 1983, pp. 80ff.; Luhmann, *Soziale Systeme*, p. 434.

4. In the sense of Charles O. Frake, 'The Ethnographic Study of Cognitive Systems', in *Anthropology and Human Behavior*, Washington D.C. 1962, pp. 72–85 (78ff.). Cf., also Frake 'The Diagnosis of Disease Among the Subanun of Mindanao', in *American Anthropologist*, vol. 63 (1961), pp. 113–32.

5. See as one example the distinction of first eternal law and second eternal law in Richard Hooker, *On the Laws of Ecclesiastical Policy*, Book 1, III, 1, cited according to the edition of Everyman's Library, Letchworth, Herts 1954, pp. 154ff.

6. This is very clear in Joyce O. Appleby, *Economic Thought and Ideology in Seventeenth-Century England*, Princeton 1978.

7. According to Joseph Glanville, *The Vanity of Dogmatizing*, London 1661, reprint Hove, Sussex 1970, p. 180, 'Nature works by an invisible hand in all things.' The origin of the metaphor of the 'invisible hand' is, as far as I know, still unclear. One could suppose that the polemics against belief in miracles and divine providence, against the 'pointing finger' of God, provided, through the occurrence of unusual events, i.e., through arguments like the ones promoted within the Royal Society, the occasion for the transformation of the metaphor of the pointing finger into that of the invisible hand. Cf., also Thomas Sprat, *The History of the Royal Society*, London 1667, pp. 82ff.

8. My proposal for this is to aim in a purely formal manner at consistency in making decisions in the legal system. Cf., Niklas Luhmann, 'Gerechtigkeit in den Rechtssystemen der modernen Gesellschaft', in *Ausdifferenzierung des Rechts: Beiträge zur Rechtssoziologie und Rechtstheorie*, Frankfurt 1981, pp. 374–418.

9. We will come back to this in chapter 16.

10. Actual discussions about dedifferentiation [*Entdifferenzierung*] and interpenetration wrestle with problems that start from the necessity of describing the process paradoxically, namely, presupposing what it supposedly eliminates. Cf., Eugen Buss/Martina Schöps, 'Die gesellschaftliche Entdifferenzierung', in *Zeitschrift für Soziologie*, vol. 8 (1979), pp. 315–29; Harald Mehlich, *Politischer Protest und Stabilität: Entdifferenzierungstendenzen in der modernen Gesellschaft*, Frankfurt 1983, especially pp. 122ff.; Richard Münch, *Theorie des Handelns: Zur Rekonstruktion der Beiträge von*

Talcott Parsons, Emile Durkheim und Max Weber, Frankfurt 1982; Münch, *Die Struktur der Moderne: Grundmuster und differentielle Gestaltung des institutionellen Aufbaus der modernen Gesellschaften*, Frankfurt 1984 (both works contain many contributions to 'interpenetration'). More circumspect and less decisive is Peter Weingart, 'Verwissenschaftlichung der Gesellschaft – Politisierung der Wissenschaft', in *Zeitschrift für Soziologie*, vol. 12 (1983), pp. 225–41.

Chapter 10   Economy

1. c.f., for more detail Niklas Luhmann, 'Das sind Preise', in *Soziale Welt*, vol. 34, (1983), pp. 153–70; Luhmann, 'Die Wirtschaft der Gesellschaft als autopoietisches System', in *Zeitschrift für Soziologie*, vol. 13 (1984), pp. 308–27.
2. In many respects one will still, of course, be able to detect 'medieval' relations of an almost universal applicability of money in developing countries. Cf., for this and for the (again medieval-like) counter-movements, Georg Elwert, 'Die Verflechtung von Produktionen: Nachgedanken zur Wirtschaftsanthropologie', in Ernst Wilhelm Müller et al. (eds), *Ethnologie als Sozialwissenschaft*, special edition 26/1984 of the *Kölner Zeitschrift für Soziologie und Sozialpsychologie*, Opladen 1984, pp. 379–402 (397ff.).
3. The modern discussion of 'property rights' has developed within this context. Its extension to ecological goods, like 'rights' to environmental pollution, has not been able to assume the former preservative function of property because a right to pollution of the air and water, whatever has been paid for it, does not enable the proprietor to handle the air or water carefully and does not give him the right to complain about the emissions of others.
4. Cf., for example, Raymond de Roover, 'The Concept of Just Price: Theory and Economic Policy', in *Journal of Economic History*, vol. 18 (1958), pp. 418–34; de Roover, *La Pensée économique des scolastiques: Doctrines et méthodes*, Paris 1971, see especially pp. 59ff.
5. This demonstrates both a release of the economy for self-regulation as well as the reinforcement of economic dependence on a functioning legal system, i.e., the increase of the dependence and independence of the economy on law. This was clearly seen by Max Weber and has been developed extensively ever since. Cf., for example, James William Hurst, *Law and the Conditions of*

Freedom in the Nineteenth-Century United States, Madison Wis. 1956. See also Hurst, *Law and Social Process in United States History*, Ann Arbor Mich. 1960; Hurst, *Law and Economic Growth: The Legal History of the Lumber Industry in Wisconsin 1836–1915*, Cambridge Mass. 1964 (with indirect references to ecological consequences); and Morton Horwitz, *The Transformation of American Law 1780–1860*, Cambridge Mass. 1977. Worthy of consideration is Warren J. Samuels, 'Interrelations between Legal and Economic Processes', in *Journal of Law and Economics*, vol. 14 (1971), pp. 435–50.

6. Thus Michel Aglietta/André Orléan, *La Violence de la monnaie*, 2nd edn, Paris 1984. Drawing on René Girard the concept is called here 'imitative contagion' and designates the interconnectedness of mimic behavior, i.e., manufactured scarcity, conflict and violence as the conditions of order.

7. This is in blatant opposition to what jurists, planners and even economists who require a 'basic order' normally think. The system no longer reacts to structural goals with conformity/deviance but only to structural changes that can be perceived and processed.

8. To be sure, very uncertain estimations! See, for example, *Handelsblatt*, of 28 February 1985; *Börsenzeitung* of 1 March 1985; *Herald Tribune* of 4 March 1985; *Frankfurter Allgemeine Zeitung* of 7 March 1985 – all of these relating to the problem of the effect on the dollar of the intervention of the central banks.

9. One finds in the 'environmental economy', as with Hans Christian Binswanger, 'Ökonomie und Ökologie – neue Dimensionen der Wirtschaftstheorie', in *Schweizerische Zeitschrift für Volkswirtschaft und Statistik*, vol. 108 (1972), pp. 251–81 (276ff.), the interpretation that the real reason for the expansion of the economy and for the increasing burdening of the environment is found in the creation of money. But just as with every ascription this is problematic because it promotes the impression that one can remedy the trouble at this point.

10. Therefore it is not without reason that one has spoken of the 'sovereignty' of money and of the ultimate arbitrariness of its violence (although these concepts suggest an analogy with politics that can be overextended and misunderstood). Cf., in connection with René Girard, Michel Aglietta/André Orléan, *La Violence de la monnaie*, Paris 1984, especially pp. 53ff.

11. *Makroökonomik des Umweltschutzes*, Göttingen 1976, p. 10.

12. One can also interpret this as the failure of a 'hierarchization' with which the economic system normally weakens its fundamental

paradox. Or, formulated differently, the environmental economy has to resort to other forms of paradox-elimination and asymmetrization.

13. Less optimistic are observers who simply begin from the facts. Cf., for example, Brock B. Bernstein, 'Ecology and Economics: Complex Systems in Changing Environments', in *Annual Review of Ecology and Systematics*, vol. 12 (1981), pp. 309–30 and in relation to the boundaries of the consequences of a 'moral suasion', of a change of values, of a change of consciousness William J. Baumol/Wallace E. Oates, *Economics, Environmental Policy and the Quality of Life*, Englewood Cliffs, N.J. 1979, pp. 282ff.

14. Cf., for example, Karl-Heinrich Hansmeyer, 'Ökologische Anforderungen an die staatliche Datensetzung für die Umweltpolitik und ihre Realisierung', in Lothar Wegehenkel (ed.), *Marktwirtschaft und Umwelt*, Tübingen 1981, pp. 6–20 (9).

15. I have not been able to understand and to translate into a sociological language what economists understand by the 'market'. The crucial systems-theoretical insight is that the market is not a 'subsystem' of the economic system but its system-internal environment or section of this environment viewed from the perspective of the individual subsystems. Cf., especially Harrison C. White, 'Where Do Markets Come From?' in *American Journal of Sociology*, vol. 87 (1981), pp. 517–47. If one begins from this then there is no difficulty in discovering such system-internal environments even in socialist economies with a state-run production apparatus. Whether one calls this a 'market' or not is then primarily a matter of ideology.

16. Reflections on theoretical models can be found in Horst Siebert, *Ökonomische Theorie der Umwelt*, Tübingen 1978.

17. Cf., Douglas R. Hofstadter, *Gödel, Escher, Bach: An Eternal Golden Braid*, Hassocks, Sussex 1979.

18. For further problems resulting from the distinction of level-decisions and allocation-decisions see also Joachim Klaus, 'Zur Frage der staatlichen Fixierung von Umweltstandards und Emissionsniveaus', in Wegehenkel, ibid., pp. 96–9.

19. Compared with the legal system, the parallel is clear with the similarly hierarchical difference of levels between law-making and law-application that has to be viewed as a strategy of paradox-elimination but which always fails in practice. This, however, occurs only in individual cases and in a way that is acceptable.

20. See perhaps Bender, ibid., (1976); Sieber, ibid., (1978); or in reference to the level of management science Udo Ernst Simonis

(ed.), *Ökonomie und Ökologie: Auswege aus einem Konflikt*, Karlsruhe 1980.

21. It is then an empirical question whether and how far these effects can be compensated by new demands. In any event, the theory does not require that increased expense for environmental considerations has to be detrimental to the entire economy.

22. In this sense it is meaningless to speak of 'non-economic' costs. This is only a metaphorical way of speaking that transfers the specificity of the economic mode of thinking indiscriminately to other social domains.

23. Thus we must reject the interpretation that can be found occasionally in the economic literature (for example, Abraham Moles/Elisabeth Rohmer, *Théorie des actes: vers une écologie des actions*, Paris 1977, p. 57), which says that through the calculations of costs a *social* integration of action in terms of scarce resources can be reached.

Chapter 11   Law

1. For the controversy cf., the detailed presentation of William J. Baumol/Wallace E. Oates, *Economics, Environmental Policy and the Quality of Life*, Englewood Cliffs, N.J. 1979, pp. 230ff.

2. For more detail cf., Niklas Luhmann, *Rechtssoziologie*, 2nd edn, Opladen 1983, pp. 354ff.; Luhmann, 'Die Einheit des Rechtssystems', in *Rechtstheorie*, vol. 14 (1983), pp. 129–54.

3. Wherever this condition is not met – in the slums of the larger cities of Brazil it is said that the people live according to their own laws, not according to the laws of the state – the certainty that the law is not illegal does not exist.

4. For the sake of clarification it should be said that, in opposition to a widespread belief (see Lawrence M. Friedman, *The Legal System: A Social Science Perspective*, New York 1975), we do not restrict the legal system to its organizational and professional working complex (law-making, judicial procedure, attorneys) but include any communication that guides itself by the difference of legal and illegal in the juristic sense.

5. That the 'complete' description of the world remains logically incomplete because third possibilities have to be excluded remains a persistently recurring problem of the legal tradition. Cf., Niklas Luhmann, 'Die Theorie der Ordnung und die natürlichen Rechts', in *Rechtshistorisches Journal*, vol. 3 (1984), pp. 133–49.

6. Correspondingly, introductory and class texts have been in production for more than ten years. See, for example, Michael Klöpfer, *Zum Umweltschutzrecht in der Bundesrepublik*, Perscha n.d. (1972); Peter-Christoph Storm, *Umweltrecht: Einführung in ein neues Rechtsgebiet*, Berlin 1980; Juergen Salzwedel (ed.), *Grundzüge des Umweltrechts*, Berlin 1982.

7. Michael Klöpfer, *Systematisierung des Umweltrechts*, Berlin 1978, serves as a good survey.

8. For this it is already of secondary importance whether particular freedom-rights, like the right to use the air, fall under constitutional freedom-guarantees or whether they are viewed as legal positions that are initially guaranteed by the law-makers and afterwards are modifiable if necessary.

9. The picture is no different if, instead of this, one has in mind the more formal concepts of permission and prohibition or indicates that the consideration of interests or the distinction of centralized and decentralized standards of selection play an important role. Even in these cases the distinction that structures the juristic discourse politically or the interpretation of the law *is not the one of system and environment*.

10. From Robert Weimar/Guido Leinig, *Die Umweltvorsorge im Rahmen der Landesplanung Nordrhein-Westfalen*, Frankfurt 1983, pp. 20 or 40.

11. Of course, there are many other reasons for the often lamented 'performance deficit'. Cf., Karl-Heinrich Hansmeyer (ed.), *Vollzugsprobleme der Umweltpolitik: Empirische Untersuchungen der Implementation von Gesetzen im Bereich der Luftreinhaltung und des Gewässerschutzes*, (project director Renate Mayntz) n.d. 1978. A good case-study is Bruce A. Ackerman et al., *The Uncertain Search for Environmental Quality*, New York 1974. In these details one immediately sees how difficult it is to present 'thoroughgoing' proposals for improvement.

12. Cf., Robert D. Luce/Howard Raiffa, *Games and Decisions*, New York 1957, especially pp. 278ff.

13. Cf., Aaron Wildavsky, 'No Risk is the Highest Risk of All', in *American Scientist*, vol. 67 (1979), pp. 32–7. Cf., also Peter Gärdenfors, 'Forecasts, Decisions and Uncertain Probabilities', in *Erkenntnis*, vol. 14 (1979), pp. 159–81, with reference to the significance of the distinctive *quality* of prognoses – a question whose omission in decisions according to the rule of the maximization of profit makes these decisions *too risky*.

14. Cf., Nathan Kogon/Michael A. Wallach, 'Risk-Taking as a Function

of the Situation, the Person, and the Group', in *New Directions in Psychology III*, New York 1967, pp. 111–278. Coming from entirely different points of departure, investigations in decision theory provide the same impression. Cf., Harry J. Otway, 'Perception and Acceptance of Environmental Risks', in *Zeitschrift für Umweltpolitik*, vol. 2 (1980), pp. 593-616 (with reference to groups that are more strongly committed) or Baruch Fischhoff et al., *Acceptable Risk*, Cambridge, UK 1981.

15. And this in the upper and lower strata, with or without calculation. Cf., for its scope, Jaques de Caillière, *La Fortune des gens de qualité et des gentilhommes particuliers*, Paris 1664, pp. 307ff.; Hunter S. Thompson, *Hell's Angels*, New York 1966.

16. For extensive research cf., David Kahneman/Paul Slovic/Amos Tversky, *Judgement under Uncertainty*, Cambridge, UK 1982.

17. Cf., for this argument Chauncey Starr/Richard Rudman/Chris Whipple, 'Philosophical Basis for Risk Analysis', in *Review of Energy*, vol. 1 (1976), pp. 629–62. A survey of (very inadequate) empirical methods for the ascertainment and collection of social preferences for risk are found in William D. Rowe, *An Anatomy of Risk*, New York 1977, pp. 259ff.

18. Thus Heinhard Steiger, 'Verfassungsrechtliche Grundlagen', in Salzwedel, ibid., pp. 21–63, for 'risk-remainder', pp. 37ff., quote p. 41.

19. Thus Hasso Hofmann, *Rechtsfragen atomarer Entsorgung*, Stuttgart 1981, for risk-remainder especially pp. 336ff. Or pluralized according to organized interests as with Karl-Heinz Ladeur, '*Abwägung' – Ein neues Paradigma des Verwaltungsrechts: Von der Einheit der Rechtsordnung zum Rechtspluralismus*, Frankfurt 1984.

20. Cf., with examples from American legislation, Talbot Page, 'A Generic View of Toxic Chemicals and Similar Risks', in *Ecology Law Quarterly*, vol. 7 (1978), pp. 207–44. Cf. also Lawrence H. Tribe, 'Trial by Mathematics: Precision and Ritual in the Legal Process', in *Harvard Law Review*, vol. 84 (1971), pp. 1329–93.

21. The history of the concept of risk is still unclear. The occasion for the appearance of a special concept distinct from the general concept of danger could also have been the need to view risks not only negatively as dangers but to consider them as the object of an intentional undertaking and the will to pay for their absorption.

22. The focus of the discussion lies in the question of the creation of marketable emissions laws. For the already extensive consideration of the advantages and disadvantages cf., the many contributions in

Lothar Wegehenkel (ed.), *Marktwirtschaft und Umwelt*, Tübingen 1981; further Werner Zohlhöfer, 'Umweltschutz in der Demokratie', in *Jahrbuch für Neue Politische Ökonomie*, vol. 3 (1984), pp. 101–21.

23. For the problem of the quantification of the readiness to take fatality risks into consideration and for the difficulty in determining ethical standards for this cf., Ronald A. Howard, 'On Making Life and Death Decisions', in Richard C. Schwing/Walter A. Albers, Jr, (eds), *Societal Risk Assessment: How Safe is Safe Enough?*, New York 1980, pp. 89–106. Obviously the problem is significant not only for technological or ecological risks. But it acquires an actuality and recognition through them.

24. Of course, even this is not true unconditionally, but only, as reactions to the Three Mile Island accident indicate, to very different degrees. Cf., Ortwin Renn, *Wahrnehmung und Akzeptanz technischer Risiken*, vol. III, Jülich 1981, pp. 20ff. For the time being there is no explanation for these distinctions.

25. Bryan Wynne, 'Redefining the Issues of Risk and Public Acceptance: The Social Viability of Technology', in *Futures*, vol. 15 (1983), pp. 13–32, makes this clear.

26. For the cybernetics of the oscillating between internal and external perspectives in the system cf., Stein Bråten, 'The Third Position: Beyond Artificial and Autopoietic Reduction', in *Kybernetes*, vol. 13 (1984), pp. 157–63.

27. That problems result from this transferring of questions that cannot be decided in a purely juristic way into the political system can be shown with a consideration of the Green parties. We will come back to this in chapter 8.

28. Cf., the widely discussed remarks of John Rawls, *A Theory of Justice*, Cambridge Mass. 1971 or Jürgen Habermas, *Theorie des kommunikativen Handelns*, 2 vols., Frankfurt 1981.

29. See the above-mentioned study of Hunter Thompson, *Hell's Angels*, New York 1966. See also Erving Goffman, 'Where the Action Is', in Goffman, *Interaction Ritual: Essays in Face-to-Face Behavior*, Chicago 1967, pp. 149–270; or for another extreme case, i.e., the climbing of the Himalayas, Michael Thompson, 'Aesthetics of Risk: Culture or Context', in Richard C. Schwing/Walter A. Albers, Jr, (eds), *Societal Risk Assessment: How Safe is Safe Enough?*, New York 1980, pp. 273–85.

30. Cf., the already out-of-date inquiries in Volkmar Gessner et al., *Umweltschutz und Rechtssoziologie*, Bielefeld 1978, pp. 167ff. Since then judicial activity has certainly increased but there are no

new investigations. From the juristic point of view see Michael Klöpfer, 'Rechtsschutz im Umweltschutz', in *Verwaltungsarchiv*, vol. 76 (1985), pp. 371–97 (Part 2 forthcoming).

31. See note 11 above.
32. Cf., Barry Boyer/Errol Meidinger, 'Privatizing Regulatory Enforcement: A Preliminary Assessment of Citizen Suits Under Federal Environmental Laws', MS Buffalo N.Y. 1985. (I owe the reference to this work to Volkmar Gessner).
33. Cf., Gerd Winter, 'Bartering Rationality in Regulation', in *Law and Society Review*, vol. 19 (1985), pp. 219–50.
34. See dispute Klöpfer ibid., (1985) p. 391 with references.

Chapter 12   Science

1. To simplify the presentation we will omit here the argument used by Heidegger, for instance, that the original difference between true and untrue had already been replaced by the difference between correct and false (especially in reference to ideas) in classical Greek philosophy and that the result was a loss of being that has not been remedied up to the present. If this reconstruction of philosophical semantics is correct, then a correlate of the beginning and increasingly social differentiation of science can be found here.
2. See for this transformation Rudolph Stichweh, *Zur Entstehung des modernen Systems wissenschaftlicher Disziplinen: Physik in Deutschland 1740–1890*, Frankfurt 1984.
3. See Edmund Husserl, *Die Krisis der europaeischen Wissenschaften und die transzendentale Phaenomenologie*, Husserliana, vol. VI, The Hague 1954.
4. Tenbruck speaks of trivialization somewhat in this sense. See Friedrich H. Tenbruck, 'Wissenschaft als Trivialisierungsprozess', in Nico Stehr/Volker Meja (eds), *Wissenschaftssoziologie: Studien und Materialen*, special edition 18 of the *Kölner Zeitschrift für Soziologie und Sozialpsychologie*, Opladen 1975, pp. 19–47.
5. Formulated in a more precise, systems-theoretical sense this means that, on the one hand, such 'noise' is an indispensable stimulus of research and continually confers on it the certainty of reality. On the other hand, it must continually be transformed into information and eliminated as disturbance. Accordingly, one can expect massive irritation from the researcher's 'change of consciousness' to 'consciousness of the environment' but not automatically the

conferral of scientific significance. Once again, this is a phenomenon of limited resonance.

6. Moreover, this insight was already possible in classical modern thought. Edward Reynolds, *A Treatise of the Passions and Faculties of the Soule of Man*, London 1640, reprint Gainesville, Fla. 1971, p. 503 says, 'It is speedier to come to a Positive Conclusion by Negative Knowledge, than a naked Ignorance.' Reynolds also includes extensive remarks about the human causes of error.

7. See, in the context of a general theory of symbolically generalized media of communication, Niklas Luhmann, 'Einführende Bemerkungen zu einer Theorie symbolisch generalisierter Kommunikationsmedien', in Luhmann, *Soziologische Aufklärung*, vol. 2, Opladen 1975, pp. 170–92.

8. Cf., in comparison to a more modern, specifically scientific rationality, Peter-Michael Spangenberg, *Maria ist immer und überall*, Frankfurt 1987.

9. Cf., for this Niklas Luhmann, 'The Differentiation of Advances in Knowledge', in Nico Stehr/Volker Meja (eds), *Society and Knowledge: Contemporary Perspectives on the Sociology of Knowledge*, New Brunswick 1984, pp. 103–48.

10. The rejection of *curiositas* was not meant to eliminate the striving for knowledge as such, only knowledge that was pointless – whether this was in reference to transcendent matters that were accessible only through faith or things that were arcane by their very nature and could only be destroyed by knowledge. Moreover, as an interest for 'others', *curiositas* was reproached for directing attention away from important *self*-knowledge. Thus the dispute was not 'innovation versus constancy'. This distinction was introduced only after the raising of *curiositas* to the universal striving for knowledge. Cf., for example, Thomas Wright, *The Passions of the Minde in General*, rev. edn, London 1630, reprint Urbana Ill., 1971, pp. 312ff.; Reynolds, ibid., (1640), pp. 462ff. And, of course, there is Hans Blumenberg, *Der Prozess der theoretischen Neugierde*, Frankfurt 1973.

11. Historically, this classification of the concept of method begins after the widespread effect of printing in the sixteenth century. The most important transitional position is occupied by Petrus Ramus (Pierre de la Rammée) who understood method as a binary schema but still applied it directly to the decomposition of reality. Cf., Walter J. Ong, *Ramus: Method and the Decay of Dialogue: From the Art of Discourse to the Art of Reason*, Cambridge Mass. 1958, reprint New York 1979.

12. Cf., for example, Isaac Levi, *Gambling with Truth: An Essay on Induction and the Aims of Science*, London 1967.

13. The theory of evolution, therefore, is at the very least, a significant perspective of modern science because it *comes to the rescue here* and explains how (not why) reality, without any consideration of logic and mathematics, *so simplified itself* that it finally has become what it is.

14. This has been formulated ever since Warren Weaver, 'Science and Complexity', in *American Scientist*, vol. 36 (1948), pp. 536–44. Cf., also Todd R. LaPorte (ed.), *Organized Social Complexity: Challenge to Politics and Policy*, Princeton N.J. 1975; Giovan Francesco Lanzara/Francesco Pardi, *L'interpretazione della complessità: Methodo sistemico e scienze sociali*, Naples 1980; Hans W. Gottinger, *Coping with Complexity*, Dordrecht 1983.

15. Cf., for example, Henri Atlan, *Entre le cristal et la fumée: Essai sur l'organisation du vivant*, Paris 1979, pp. 74ff. Cf., also Lars Löfgren, 'Complexity Descriptions of Systems: A Foundational Study', in *International Journal of General Systems*, vol. 3 (1977), pp. 197–214; Robert Rosen, 'Complexity as a System Property', in *International Journal of General Systems*, vol. 3 (1977), pp. 227–32.

16. Cf., Heinz von Foerster, *Observing Systems*, Seaside Ca. 1981, especially pp. 288ff.

17. It is obvious that this is an offense against the rules of the theory of science – even against the rules of the divine Popper. See, for example, Hans Albert, 'Modell-Platonismus: der neoklassische Stil des ökonomischen Denkens in kritischer Beleuchtung', in *Festschrift Gerhard Weisser*, Berlin 1963, pp. 45–76. But this insight does not lead one out of the problem but deeper into it when it raises the question (and this is only a different version of the problem of structured complexity) how one can protect the theory of science against infection by paradoxes, how one can immunize Popper himself.

18. A quite considerable investigation of the effects of distinctions of size on social structures is summarized as follows, '(25) Other things being equal, the above statements about the relationship between scale and social organizations are true. (26) Other things are never equal.' Gerald D. Berreman, 'Scale and Social Relations: Thoughts and Three Examples', in Frederick Barth (ed.), *Scale and Social Organization*, Oslo 1978, pp. 41–77 (77).

19. To be sure, not if one considers the relative isolation of life on the earth. Measured by the analytic apparatus of scientific theories

this boundary is not an 'ecological' boundary any longer. Moreover, the totality of events on the earth is much too complex for anyone to be able to work with this system reference scientifically.

20. Thus, for example, Roy A. Rappaport, *Ecology, Meaning, and Religion*, Richmond Ca. 1979, pp. 54ff.

21. Cf., Herbert Simon/Albert Ando, 'Aggregation of Variables in Dynamic Systems', in *Econometrica*, vol. 29 (1961), pp. 111–38; Herbert A. Simon, 'The Architecture of Complexity', in *Proceedings of the American Philosophical Society*, vol. 106 (1962), pp. 467–82 (reprinted elsewhere); Franklin M. Fisher/Albert Ando, 'Two Theorems on *Ceteris Paribus* in the Analysis of Dynamic Systems', in *American Political Science Review*, vol. 56 (1962), pp. 108–33; Albert Ando/Franklin M. Fisher, 'Near Decomposability, Partition and Aggregation and the Relevance of Stability Discussions', in *International Economic Review*, vol. 4 (1963), pp. 53–67; Albert Ando/Franklin M. Fisher/Herbert A. Simon, *Essays on the Structure of Social Science Models*, Cambridge Mass. 1963. Cf., further C. West Churchman, *The Design of Inquiring Systems: Basic Concepts of System and Organization*, New York 1971, especially pp. 64ff.; Daniel Metlay, 'On Studying the Future Behavior of Complex Systems', in LaPorte, ibid., (1975), pp. 220–50; William C. Wimsatt, 'Complexity and Organization', in Marjorie Green/ Everett Mendelsohn (eds), *Topics in the Philosophy of Biology*, Boston Studies in the Philosophy of Science, 27 (1976), pp. 174–93. As the contributions quoted last indicate, the sceptical estimation is increasing.

22. And if unattainability is a foregone conclusion then it is pointless to look for rationality in the establishment and the adherence to procedural rules for reaching rationality – the 'bourgeois' way of proceduralizing difficult questions.

Chapter 13 Politics

1. 'Politique et société', in *Communications*, vol. 22 (1974), pp. 119–33 (125).

2. Cf., for the widespread, earlier terminology that prevailed until the beginning of the eighteenth century, Daniel de Priezac, *Discours politiques*, 2nd edn, Paris 1666; Rémond de Cours, *La véritable Politique des personnes de qualité*, Paris 1692; Christian Thomasius, *Kurtzer Entwurff der politischen Klugheit*, German trans. Frankfurt-Leipzig 1710, reprint Frankfurt 1971; Jürgen Habermas,

Kleine Politische Schriften, Frankfurt 1981. In this context politics means not much more than public behavior.

3. For example, Ciro Spontone, *Dodici libri del governo di Stato*, Verona 1599.

4. We are talking here in the context of the systems theory of social systems, i.e., about communication. That the situation is different for psychical systems is obvious.

5. Cf., Niklas Luhmann, 'Der politische Code: "Konservativ' und "progressiv" in systemtheoretischer Sicht', in Luhmann, *Soziologische Aufklärung*, vol. 3, Opladen 1981, pp. 267–86; Luhmann, *Politische Theorie im Wohlfahrtsstaat*, Munich 1981, pp. 118ff.

6. This argument can be supported by numerous other comparisons. To simplify the presentation we have restricted ourselves to the differentiation of coding and programming that is relevant for the particular theme of resonance capacity.

7. Cf., Manfred Schmitz, *Theorie und Praxis des politischen Skandals*, Frankfurt 1981; Francesco M. Battisti, *Sociologia dello scandalo*, Bari 1982. An empirical investigation of historical scandals would supposedly show quite readily that ecological interests also increase in this form, i.e., have become capable of being scandalous – either because the total number of scandals increases or because the distribution within this total shifts from morality to ecology. The moralization of ecological themes may then no longer have the function of making them capable of becoming scandalous.

8. For more on this cf. Niklas Luhmann, *Macht*, Stuttgart 1975, pp. 60ff.

9. We will come back to this later, pp. 283ff.

10. This case is exactly parallel, in a systems-theoretical sense, to the question that we have been pursuing with respect to society. In both cases we are concerned with whether and how a system can find its rationality through calculating the effects of its own operations on its environment in reference to the reactions on itself.

11. Cf., for this also Walter Bühl, 'Ökologische Knappheit', in ibid., pp. 141ff.

12. See especially Peter Graf Kielmansegg, 'Politik in der Sackgasse? Umweltschutz in der Wettbewerbsdemokratie', in Heiner Geissler (ed.), *Optionen auf eine lebenswerte Zukunft: Analysen und Beiträge zur Umwelt und Wachstum*, Munich 1979, pp. 37–56.

Chapter 14 Religion

1. In Horst Westmüller, 'Die Umweltkrise – eine Anfrage an Theologie und Christen', in Hans Dietrich Engelhardt (ed.), *Umweltstrategie: Materialien und Analysen zu einer Umweltethik der Industriegesellschaft*, Gütersloh 1975, pp. 314–48, 331). The quotation is chosen arbitrarily and ought not to slander its author, but it is representative for what I have found everywhere in the literature about the religious position towards the environmental crisis. Cf., also Martin Rock, 'Theologie der Natur und ihre anthropologischen Konsequenzen', in Dieter Birnbacher (ed.), *Ökologie und Ethik,*, Stuttgart 1980, pp. 72–102; and Gerhard Liedke, *Im Bauch des Fisches: Ökologische Theologie*, Stuttgart 1979, with a much more direct biblical orientation.
2. Formulated in terms of complexity one could also say its ultimate difference resides in the indeterminability (transcendence) of the determinate (immanence). This is the formulation that I use in *Die Funktion der Religion*, Frankfurt 1977.
3. See also the factor of 'surprise' in Matthew 25, 31ff. as an expression of the difference of immanent and transcendent valuation.
4. Or as Shaftesbury had already dared, ibid., vol. III, p. 316 and repeatedly dared to say that faith is irreproachable, 'as by Law Established'.
5. Johann Heinrich Lambert, *Cosmologische Briefe über die Einrichtung des Weltbaues*, Augsburg 1761, p. 116.
6. Anthropological formulas for this – like the self-created inscrutability of God (fear of humanity), the cunning of God, the screening of humanity from unbearable knowledge – are certainly old. Cf., some material in Stephen D. Benin, 'The "Cunning of God" and Divine Accommodation', in *Journal of the History of Ideas*, vol. 45 (1984), pp. 179–91.
7. With this in mind I decided to use the Montaigne quote as a motto. It ought to be considered in that context.
8. Michel Serres, *Le Parasite*, Paris 1980.

Chapter 15 Education

1. Cf., Norman J. Faramelli, 'Ecological Responsibility and Economic Justice', in Ian G. Barbour (ed.), *Western Man and Environmental*

Ethics: Attitudes Toward Nature and Technology, Reading Mass. 1973, pp. 188–203.

2. For this concept of career cf., Niklas Luhmann/Karl Eberhard Schorr, *Reflexionsprobleme im Erziehungssystem*, Stuttgart 1979, pp. 277ff.

3. This estimation is also found in Heinz von Foerster, *Observing Systems*, Seaside Ca. 1981, pp. 209ff. – concluding with the proposal, 'Would it not be fascinating to think of an education system that detrivializes its students by teaching them to ask "legitimate questions," that is, questions for which the answers are unknown?'

4. Just to mention others: the problem of observing and understanding in a system with structured complexity; the problem of actor/observer differences of attribution; the problem of the 'hidden agenda' and the socialization for survival in school.

5. So suppose William J. Baumol/Wallace E. Oates, *Economics, Environmental Policy, and the Quality of life*, Englewood Cliffs N.J. 1979, pp. 282ff. (although their own research results, concerning recycling, seem to contradict this).

Chapter 16   Functional Differentiation

1. Cf., among others Jeffrey L. Pressman/Aaron Wildavsky, *Implementation: How Great Expectations in Washington are Dashed in Oakland*, Berkeley Ca. 1973.

2. Cf., Niklas Luhmann, *Soziale Systeme*, Frankfurt 1984, pp. 37ff.

3. According to E. T. A. Hoffmann, 'Des Vetters Eckfenster', *Werke*, Berlin-Leipzig, no date, vol. 12, pp. 142–64.

4. Cf., for a historico-semantic context Niklas Luhmann, 'Temporalisierung von Komplexitaet: Zur Semantik neuzeitlicher Zeitbegriffe', in Luhmann, *Gesellschaftsstruktur und Semantik*, vol. 1, Frankfurt 1980, pp. 235–300.

5. Cf., André Béjin, 'Différenciation, complexification, évolution des sociétés', in *Communications*, vol. 22 (1974), pp. 109–18 (114) in connection with Henri Atlan, *L'Organisation biologique et la théorie de l'information*, Paris 1972, pp. 270ff.

6. The still unclear semantic career of the concept of value (especially prior to the middle of the nineteenth century) might have one of its sources here. To be sure, it is incorrect to say that the concept of value was appropriated by morality, literature, aesthetics and philosophy from economics only in the middle of the nineteenth

century. (The Abbé Morellet, *Prospectus d'un nouveau dictionnaire de commerce*, Paris 1769, reprint Munich 1980, pp. 98ff., observes a restriction to economic profit. But the entire eighteenth century used it in a much more general sense). It is equally clear, however, that the concept of value has been used as an ultimate guarantee for meaning and therefore non-contradictably in the last hundred years.

7. This happens in any event. But it is also required in many respects and viewed as the precondition for the solutions of problems. Cf., Karl-Heinz Hillmann, *Umweltkrise und Wertwandel: Die Umwertung der Werte als Strategie des Überlebens*, Frankfurt-Bern 1981.

8. Cf., Talcott Parsons, 'On the Concept of Value-Commitments', in *Sociological Inquiry*, vol. 38 (1968), pp. 135–60 (153ff.).

9. For a comparison: the self-description of stratified societies had always used a moral schematism – whether in the direct moral criticism of typical behavior in the individual strata or in the formulation of types of perfection from which everyone could measure their distance.

10. Cf., for example, Simon-Nicolas-Henri Linquet, *Le Fanatisme des philosophes*, London-Abbeville 1764; Peter Villaume, *Über das Verhältnis der Religion zur Moral und zum Staate*, Libau 1791, and of course, the widespread critique of the French Revolution as the outbreak of a naive faith in principles.

11. This led many to the conclusion of 'revolution' – with very little support for possibilities and consequences. One finds typically that the manifest/latent schema is introduced without further reflection as a description of facts and forms the basis for analyses. This has been the case especially since Robert K. Merton, 'The Unanticipated Consequences of Purposive Social Action', in *American Sociological Review*, vol. 1, (1936), pp. 894–904.

12. Jürgen Habermas judges much more sharply and leaves more room for hope. He views this as the theory-immanent *problem* of the Enlightenment's erroneous semantic guidance by the theory of the subject and its object and therefore sees the *solution* of the problem in the transition to a new paradigm of intersubjective agreement. Cf., *Der philosophische Diskurs der Moderne: Zwölf Vorlesungen*, Frankfurt 1985. To make this useful sociologically, one must still clarify how this erroneous guidance and the possibility of correcting it are connected with the structure of modern society.

Chapter 17   Restriction and Amplification

1. Thus according to Simon-Nicolas-Henri Linguet, 'Lettre sur *La Théorie des loix civiles*', Amsterdam 1770, p. 96 (in connection with a critique of Montesquieu). To earlier authors who began from the aristotelian theory of motion equilibrium was seen as corruption, i.e., as indecisiveness in the direction of motion – 'as it were in aequilibrio, that it cannot tell which way to encline', as Reynolds says, ibid., (1640), p. 463.

Chapter 18   Representation and Self-observation

1. We will leave aside the special case of Egypt in which neither the one nor the other is true but, instead, where religion assumed the representation of unity. It found no successors.
2. Marcel Gauchet, 'L'Expérience totalitaire et la pensée de la politique', in *Esprit*, July/August 1976, pp. 3–28 (26).
3. For potentializing through inhibition cf., Yves Barel, *Le Paradoxe et le système: Essai sur la fantastique social*, Grenoble 1979, pp. 185ff.
4. The recursivity in the formulation is intended.
5. The impetus that the theory of attribution experienced within this context was itself an interesting theme of research. It is still visible, for example, in Felix Kaufmann, *Methodenlehre der Sozialwissenschaften*, Wien 1936, especially pp. 181ff.
6. For a topical survey cf., Ortwin Renn, 'Die alternative Bewegung: Eine historisch-soziologische Analyse des Protestes gegen die Industriegesellschaft', in *Zeitschrift für Politik*, vol. 32 (1985), pp. 153–94; Karl-Werner Brand (ed.), *Neue soziale Bewegungen in Westeuropa und den USA: Ein internationaler Vergleich*, Frankfurt 1985.
7. In the sense of Michel Serres, *Le Parasite*, Paris 1980.

Chapter 19   Anxiety, Morality and Theory

1. Since we are concerned here only with communication we will omit the components of emotional agitation in what follows and deal only with (the expression of) worry. For the distinction of both these components of the (psychological) concept of anxiety cf., Ralf Schwarzer, *Stress, Angst und Hilflosigkeit: Die Bedeutung*

von Kognitionen und Emotionen bei der Regulation von Belastungssituationen, Stuttgart 1981, pp. 87ff.; Schwarzer, 'Worry and Emotionality as Separate Components of Test Anxiety', in *International Review of Applied Psychology*, vol. 33 (1984), pp. 205–20. The distinction has been worked out in the so-called Test Anxiety Research. Further contributions to this can be found in the annals *Advances in Test Anxiety Research* (from 1982).

2. Cf., Anthony, Earl of Shaftesbury, *Characteristicks of Men, Manners, Opinions, Times*, 2nd edn, ibid.; no date 1714, reprint Farnborough, Hants UK 1968, vol. 1, p. 16. The collapse of a theoretical construction of a Hobbesian type can easily be seen in this.

3. See, for someone with this intention, for example, William W. Lowrance, *Science and the Determination of Safety*, Los Altos Ca. 1976.

4. Cf., as case-studies Dorothy Nelkin, 'The Role of Experts on a Nuclear Siting Controversy', in *Bulletin of the Atomic Scientists*, vol. 30 (1974), pp. 29–36; Helga Nowotny, *Kernenergie: Gefahr oder Notwendigkeit: Anatomie eines Konflikts*, Frankfurt 1979; furthermore, writings from the extensive literature Dorothy Nelkin/ Michael Pollak, 'The Politics of Participation and the Nuclear Debate in Sweden, the Netherlands and Austria', in *Public Policy*, vol. 25 (1977), pp. 333–57; Edgar Michael Wenz (ed.), *Wissenschaftsgerichtshöfe: Mittler zwischen Wissenschaft, Politik und Gesellschaft*, Frankfurt 1983.

5. A good indicator is that high grades in school among those with a positive academic attitude can accompany greater uncertainty about self-value and greater fears over performance than average grades. Cf., Helmut Fend, 'Selbstbezogene Kognition und institutionelle Bewertungsprozesse im Bildungswesen: Verschonen schulische Bewertungsprozesse den "Kern der Persönlichkeit?"', in *Zeitschrift für Sozialisationsforschung und Erziehungssoziologie*, vol. 4 (1984), pp. 251–70. In view of contrary research results, above all in regard to the inhibiting effects of anxiety on performance (cf., Schwarzer, ibid., 1981, pp. 100ff.), and the discovery that this is stronger among those who are more intelligent (cf., Henk M. van der Ploeg, 'Worry, Emotionality, Intelligence, and Academic Performance in Male and Female Dutch Secondary School Children', in *Advances in Test Anxiety Research*, vol. 3 1984, pp. 201–10), the research has to view this question as still open.

6. Cf., Heinz von Foerster, *Observing Systems*, Seaside Ca. 1981,

especially the contribution 'Objects: Tokens for (Eigen-) Behavior', pp. 274ff.

7. In any event, empirical investigations that have pursued this question have led to inconsistent results. Cf., Kenneth L. Higbeen, 'Fifteen Years of Fear Arousal: Research on Threat Appeals 1953–1968', in *Psychological Bulletin*, vol. 72 (1969), pp. 426–44; Werner D. Froehlich, 'Perspektiven der Angstforschung', in *Enzyklopädie der Psychologie*, C IV, vol. 2: 'Psychologie der Motivation', ed. Hans Thomas, Göttingen 1983, pp. 110–320 (178ff.).

8. Cf., Franz L. Neumann, *Angst und Politik*, Tübingen 1954.

9. Werner Froehlich, *Angst: Gefahrensignale und ihre psychologische Bedeutung*, Munich 1982, p. 27.

10. In any event, this is the way William C. Clark sees it in 'Witches, Floods, and Wonder-Drugs: Historical Perspectives on Risk Management', in Richard C. Schwing/Walter A. Albers, Jr (eds), *Societal Risk Assessment: How Safe is Safe Enough?*, New York 1980, ppp. 287–313. The same thing seems to be the case for other domains of anxiety, for example, for anxiety over examinations. See the results in D. Gertmann et al., 'Erste Ergebnisse einer Fragebogenuntersuchung zur Prüfungsvorberei- tung im Fach Psychologie', in Brigitte Eckstein (ed.), *Hochschulprü- fung: Rückmeldung oder Repression*, Hamburg 1971, pp. 54–9.

11. Cf., for a survey William D. Rowe, *An Anatomy of Risk*, New York 1977, pp. 119ff., 300ff.; for a (contested) quantitative estimation Chauncey Starr, 'Social Benefit versus Technological Risk: What is Our Society Willing to Pay For', in *Science*, vol. 168 (1969), pp. 1232–8. This 'double-standard' hypothesis seems to hold empirically even if one considers the inclusion of other factors. Cf., Paul Slovic/Baruch Fischhoff/Sarah Lichtenstein, 'Facts and Fears: Understanding Perceived Risk', in Schwing/Albers, ibid., pp. 181–214 (196, 205ff.).

12. For change and adequacy in the perception of risk from the political perspective cf., Meinholf Dierkes, 'Perzeption und Akzeptanz technologischer Risiken und die Entwicklung neuer Konsenssstrate- gien', in Jürgen von Kreudener/Klaus von Schubert (eds), *Tech- nikfolgen und sozialer Wandel: Zur politischen Steuerbarkeit der Technik*, Köln 1981, pp. 125–41; also, with a detailed literature and empirical studies of its own, Ortwin Renn, *Wahrnehmung und Akzeptanz technischer Risiken*, 6 vols., Jülich 1981.

13. For the fear of atomic disasters it is remarkable that this is consciously estimated counter-inductively, i.e., that the calculation

of the risk is not obtained from statistics of past accidents but is, so to say, projected freely and unrestrictedly. In contrast to the perception of other risks this is quite clear in Slovic et al., ibid., (1980), p. 193.

14. Moreover, it is remarkable here that the fears increase while the dangers clearly decrease. This is a case of the self-inducement of anxiety-related communication treated above. In this field of anxiety the 'double-standard' of involuntary/voluntary can be observed easily. One is more afraid of the chemistry of the food industry than one's own poor eating habits when in reality there is more reason for worrying about the latter.

15. In the sense of the Allport school. Cf., especially Richard L. Schanck, *A Study of a Community and its Groups and Institutions Conceived of as Behaviors of Individuals*, Princeton N.J. 1932; Ragnar Rommetveit, *Social Norms and Roles: Explorations in the Psychology of Enduring Social Pressures with Empirical Contributions from Inquiries into Religious Attitudes and Sex Roles of Adolescents from Some Districts in Western Norway*, Oslo 1955.

16. For the morality and logic of warning, cf., Lars Clausen/Wolf R. Dombrowsky, 'Warnpraxis und Warnlogik', in *Zeitschrift für Soziologie*, vol. 13 (1984), pp. 293–307.

17. Cf., Niklas Luhmann, *Soziale Systeme*, Frankfurt 1984, pp. 638ff.

18. See also Niklas Luhmann, 'The Self-Description of Society: Crisis Fashion and Sociological Theory', in *International Journal of Comparative Sociology*, vol. 25 (1984), pp. 59–72.

Chapter 20   Toward a Rationality of Ecological Communication

1. Jürgen Habermas, *Der philosophische Diskurs der Moderne: Zwölf Vorlesungen*, Frankfurt 1985, p. 432.

2. One can object that then the question is no longer the same, but this would only change the discussion to the problem of the criteria for the rationality of a problematic.

3. Cf., Douglas R. Hofstadter, *Gödel, Escher, Bach: An Eternal Golden Braid*, Hassocks, Sussex 1979. Furthermore see above chapter 5.

4. This is in accordance with Habermas in *Der philosophische Diskurs der Moderne*, Frankfurt 1985, pp. 426ff.

5. Cf., also Niklas Luhmann, *Soziale Systeme*, Frankfurt 1984, pp. 638ff.

6. According to Heinz von Foerster, *Observing Systems*, Seaside Ca. 1981.

Chapter 21 Environmental Ethics

1. For more detail see Niklas Luhmann, 'Soziologie der Moral', in Niklas Luhmann/Stephan H. Pfürtner (eds), *Theorietechnik und Moral*, Frankfurt 1978, pp. 8–116; Luhmann, 'I fondamenti sociali della morale', in Niklas Luhmann et al., *Etica e Politica: Riflessioni sulla crisi del rapporto fra società e morale*, Milan 1984, pp. 9–20.

2. In the seventeenth century this was quite explicit (and moreover a requirement projected even on God: God hates the sinner!). Cf., for example, Edward Reynoldes, *A Treatise of the Passions and Faculties of the Soule of Man*, London 1640, reprint Gainesville Fla. 1971, pp. 111ff., 137ff. For such a theory, one that no longer reflects the paradox, it is then a mere appearance (the world after the fall from grace), that both love as well as hate can have good as well as evil consequences. One could even talk of a parallel coding of morality and the passions that merely has to reckon with the fact that this can run askew.

3. Robert Musil, *Der Mann ohne Eigenschaften*, Hamburg 1952, p. 1204.

4. Only the supposition is noted that a sufficiently radical reflection has to lead consistently to the reinclusion of the paradox, perhaps in the form of justification through unjustifiability which as a generalized unjustifiability affects every critic simply because he or she participates in ethical discussions and thus, at least, recognizes the desirability of ethical justifications. Karl-Otto Apel seems to argue in the same way. Surely it is insufficient to delegate the question of justification simply to the ethical discourse alone and then to wait and see whether it comes to any results and which ones.

Index

Index by Isobel K. McLean